Chemometrics: Theory and Application

Bruce R. Kowalski, EDITOR
University of Washington

A symposium sponsored by the
Division of Computers in Chemistry
at the 172nd Meeting of the
American Chemical Society,
San Francisco, Calif.,
Sept. 2, 1976.

ACS SYMPOSIUM SERIES **52**

AMERICAN CHEMICAL SOCIETY
WASHINGTON, D. C. 1977

Library of Congress CIP Data

Chemometrics.

(ACS symposium series; 52 ISSN 0097-6156)

Includes bibliographical references and index.

1. Chemistry—Data processing—Congresses.

I. Kowalski, Bruce R., 1942– II. American Chemical Society. Division of Computers in Chemistry. III. Series: American Chemical Society. ACS symposium series; 52.

QD39.3.E46C48 542'.8 77-9088
ISBN 0-8412-0379-2 ACSMC8 52 1-288

PRINTED IN THE UNITED STATES OF AMERICA

ACS Symposium Series

Robert F. Gould, *Editor*

FOREWORD

The ACS SYMPOSIUM SERIES was founded in 1974 to provide a medium for publishing symposia quickly in book form. The format of the SERIES parallels that of the continuing ADVANCES IN CHEMISTRY SERIES except that in order to save time the papers are not typeset but are reproduced as they are submitted by the authors in camera-ready form. As a further means of saving time, the papers are not edited or reviewed except by the symposium chairman, who becomes editor of the book. Papers published in the ACS SYMPOSIUM SERIES are original contributions not published elsewhere in whole or major part and include reports of research as well as reviews since symposia may embrace both types of presentation.

CONTENTS

PREFACE

During the mid-1800s a number of events led scientists to seek a relationship among the chemical elements that were known at the time. About five years before Mendeleev's publication of what has been called the first periodic table, John Newlands published his periodic table which was not accepted by the scientific community, and was not to be recognized as an achievement until the Royal Society belatedly awarded him the Davy Medal five years after it had similarly honored Mendeleev. The remarkable discovery of Newlands was the repetitive pattern of properties of the elements when they were arranged according to Cannizzaro's new atomic weights. Thus Newlands, studying a collection of objects (elements) via the properties of each object, applied unsupervised pattern recognition to a problem of multivariate analysis long before computers and pattern recognition were invented. His approach to the discovery of the periodicity of the elements would make him one of the early chemometricians, if not the first.

Modern chemistry, as a physical science, studies chemical systems by obtaining information through the use of a variety of measurement systems. On the whole, the measurement systems available to chemists are quite sophisticated and generate data that are accurate and precise. Psychology, as a social science, studies human systems, also by making measurements. However, the data generated by psychology's measurement techniques are comparatively imprecise and inaccurate, and sometimes even nonmetric in nature. As a result of this problem experimental psychologists are eager to discover new mathematical and statistical methods to extract useful information from their observations. The area of psychology concerned with the design of experiments and the interpretation of observations is called psychometrics, and the journal *Psychometrika* has been published since 1936.

Psychology is not the only science formally searching for better methods of information extraction. Biometrics and econometrics are formal areas of study in biology and economics. In June 1974 the Chemometrics Society was founded in Seattle, Washington during an informal gathering of chemists. In a published letter to prospective chemometricians (*Journal of Chemical Information and Computer Sciences* (1975) **15**, 201), chemometrics is defined as the development and application of mathematical and statistical methods to extract useful chemical information from chemical measurements. Modern chemistry has ventured outside the controlled environment of the laboratory to tackle difficult problems with chemical measurements. This, combined with the proliferation of computers in chemical laboratories, has prompted a demand for new and improved methods to design and control experiments and to analyze the wealth of data that can be generated.

"Chemometrics: Theory and Application" represents a sampling of the work of chemometricians and does not constitute an all-inclusive review of that field. With one major exception and some minor ones this volume represents the content of a symposium under the same title presented by the Division of Computers in Chemistry at the 172nd National Meeting of the American Chemical Society, August 29 to September 3, 1976 in San Francisco. The major exception is the inclusion of a contributon by Sjöström and Wold. S. Wold was an invited speaker who unfortunately was unable to attend the meeting. The minor exceptions amount to differences in scope and emphasis between the papers delivered at the San Francisco meeting and those found in this book.

As data become easier and less expensive to acquire, there is little doubt that the chemist will be forced to rely more heavily on the computer and new mathematical and statistical analysis methods. Likewise, as instruments become more complex with multiple outputs as well as multiple inputs, the computer will assume a greater role in instrument control as well as such tasks as fault detection and even fault correction. The works of the authors found in this book clearly demonstrate that chemists are indeed interested and active in the search for better measurement system control and optimization and measurement analysis methods.

University of Washington
Seattle, Washington
March 7, 1977

BRUCE R. KOWALSKI

1

Advances in the Application of Optimization Methodology in Chemistry

STANLEY N. DEMING
Department of Chemistry, University of Houston, Houston, TX 77004

STEPHEN L. MORGAN
Department of Chemistry, University of South Carolina, Columbia, SC 29208

Many chemical measurement processes can be viewed as _systems_ (1) consisting of _inputs_, _transforms_, and _outputs_ (see Figure 1). The _primary input_ to a chemical measurement process is a _sample_, some _property_ of which is to be assigned a _numerical_ value (2). Examples of specific properties that might be measured are the percentage of iron in an ore, the concentration of calcium in a patient's blood serum, and the parts per million of hydrocarbons in urban air.

In addition to the primary input, many secondary inputs (or _factors_) can have an effect upon the numerical value _that is_ eventually assigned to the property of interest. These additional factors include temperature, reagent amount, wavelength, time, homogeneity, and the presence of interfering substances. If the numerical result of the measurement process is to be a precise representation of the property of interest, it is clearly important that the more significant of these factors must be controlled. As Mandel has stated, "The development of a method of measurement is to a large extent the discovery of the most important environmental factors and the setting of tolerances for the variation of each one of them" (3,4). Ideally, the method should be "rugged" against uncontrolled changes in the environmental factors so that the tolerances can be broad.

It is often convenient to classify factors as _known_ or _unknown_, and _controlled_ or _uncontrolled_. A further classification results if it is noted at what point a factor enters the measurement scheme (see Figure 2); specifically, is the factor associated with the measurement process itself (_e.g._, temperature, reagent amount) or is it instead associated with the sample (_e.g._, homogeneity, presence of interfering substances)?

Figure 1. Systems view of the measurement process

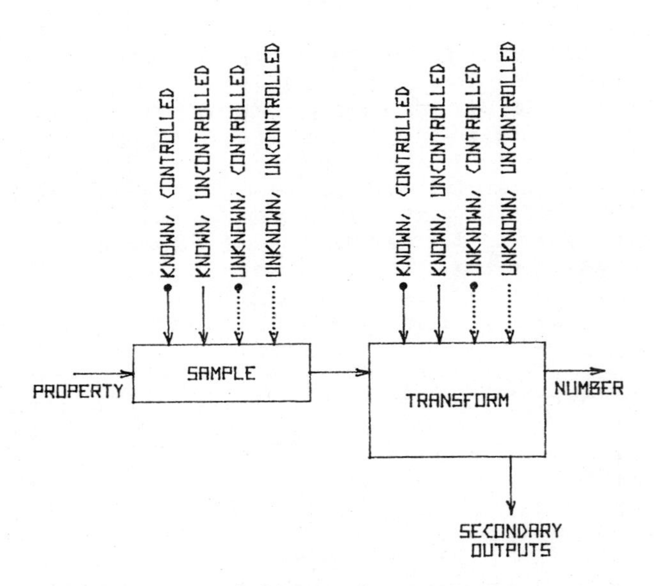

Figure 2. Expanded view of the measurement process

This latter distinction is often important in the development of analytical methods. If a known environmental factor is associated with the measurement process itself, then it is usually possible to control that factor during both the development and implementation of the method; thus, by sufficiently close control, the factor's influence on the numerical result can be minimized. If a known environmental factor is associated with the sample, it might not be possible to control that factor when the method is actually implemented. It is, however, usually possible to control the factor during the development of the method in such a way that the range of values normally encountered for that factor can be simulated. By this mechanism, the effect of a factor associated with the sample can be assessed, and the method can be developed so as to minimize the effect of this normally uncontrolled factor.

The primary output from a chemical measurement process is the numerical value of the property of interest in the sample. But many other, secondary outputs (or responses) might also be important: examples include cost per measurement, sensitivity to interfering substances, and linearity of the assigned numerical value vs. the property being measured. Thus, the development of a method of measurement can be more than the discovery of the most important environmental factors and the setting of tolerances for the variation of each one of them; it can also be the adjustment or optimization of the most important controllable environmental factors so as to achieve the best possible compromise among the many different responses (5).

The "advances" reported here illustrate the use of classical experimental designs in conjunction with optimization techniques to automatically produce a chemical measurement process possessing desirable performance characteristics (6).

Automated Continuous Flow System

Automated continuous flow methods of chemical analysis (7) have become widely accepted as reliable means of carrying out a large number of determinations in a short period of time with minimal analyst interaction. In the future, many existing continuous flow methods will need to be improved and many new continuous flow methods will need to be developed both to meet the more exacting requirements of established disciplines, as well as to fulfill the growing demands of relatively new areas such as environmental and clinical chemistries (8).

The instrument used in this work is built around standard Technicon AutoAnalyzer-II continuous flow components and a Hewlett-Packard 9830A computer. Many of the operations normally carried out by the analyst are under direct computer control. These operations include starting and stopping a tray of samples, acquiring digitized absorbance values from the colorimeter, and controlling the flow rate of individual reagents. This latter operation is accomplished by using individual peristaltic pumps for each reagent line; each peristaltic pump is driven by a stepping motor which can be made to turn at a rate that will deliver the desired flow. Computer options include: 16K bytes of read/write memory; thermal page printer; plotter; dual-platter disc; and read-only-memories for input/output, matrix, and string operations, and for advanced programming capability. A 32-bit serial, bidirectional, time multiplexed interface is used to communicate information between the instrument and computer.

Chemical System

The concentration of calcium in blood serum can be determined by dialysis of calcium ion into a recipient stream followed by reaction with the complexing agent cresolphthalein complexone in basic solution (9). Figure 3 is a diagram of the flow scheme used in this work.

Before dialysis, the serum sample is mixed with a solution containing hydrochloric acid (HCL-B), 8-hydroxyquinoline (8HQ-B), and water (used as a diluent to make up a fixed total flow). During dialysis, the calcium is transferred to a recipient stream containing hydrochloric acid (HCL-A), 8-hydroxyquinoline (8HQ-A), cresolphthalein complexone (CPC), and water. Diethylamine (DEA) is added to make the solution basic and the absorbance of the colored product is measured at 570 nm.

Figure 4 is a systems view of the continuous flow method for calcium. Six controllable factors associated with the measurement process have an influence either upon the number that is assigned to the calcium concentration, or upon some of the secondary outputs, or both. These factors are HCL-B, 8HQ-B, HCL-A, 8HQ-A, CPC, and DEA. Two uncontrollable factors that are associated with the sample are the concentrations of magnesium and protein in the serum. Magnesium is

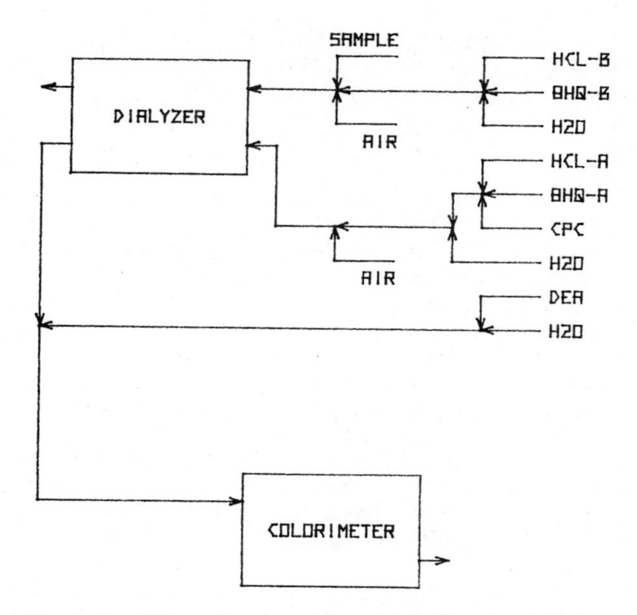

Figure 3. Flow scheme for continuous flow determination of calcium in blood serum

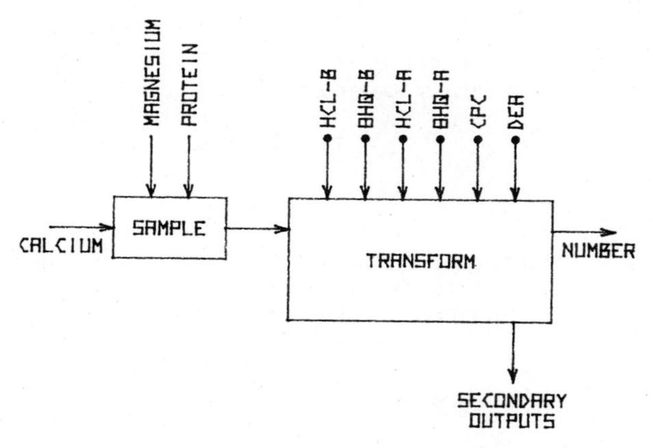

Figure 4. Systems view of continuous flow determination of calcium in blood serum

an interfering factor because it can codialyze with calcium and contribute to the measured absorbance by forming a colored Mg-CPC complex (10). Protein is an interferent because it is thought to contribute to a Donnan-type equilibrium in the dialysis process (11) and because it chemically binds calcium.

Objectives

The objectives of the work presented here were to

- -- increase the sensitivity of absorbance response with respect to serum calcium concentration (that is, maximize the slope $\delta A/\delta[CA]$);
- -- decrease the sensitivity of absorbance response with respect to serum magnesium concentration (*i.e.* minimize $|\delta A/\delta[Mg]|$);
- -- decrease the sensitivity of absorbance response with respect to serum protein concentration (minimize $|\delta A/\delta[\text{protein}]|$);
- -- maintain a relatively low blank absorbance (in this work, the blank absorbance was considered acceptably low if it was less than the absorbance above the reagent blank produced by a 15 mg dl^{-1} Ca standard); and
- -- maintain good linearity of absorbance with respect to serum calcium concentration (in this work, the standard deviation of residuals from a model first order in calcium was used to assess linearity).

Samples

A set of 20 serum samples were prepared by diluting one part reference serum with one part saline solution containing a specified amount of calcium ion, magnesium ion, and bovine serum albumin (protein). Each serum sample thus prepared corresponded to a treatment combination in the non-central composite experimental design (12) given in Table I.

Objective Function

Calcium, magnesium, and protein effects were assessed by fitting a full second-order polynomial model to data obtained from the set of 20 samples:

$$A = \beta_0 + \sum_{i=1}^{3} \beta_i x_i + \sum_{i=1}^{3} \sum_{j=1}^{3} \beta_{ij} x_i x_j \qquad (1)$$

where A is the absorbance above the reagent blank, and x_1, x_2, and x_3 correspond to calcium, magnesium, and

Table I

Experimental Design for Samples

Sample	Calcium[a]	Magnesium[a]	Protein[a]
7	-1	-1	-1
3	-1	-1	+1
14	-1	+1	-1
6	-1	+1	+1
4	+1	-1	-1
8	+1	-1	+1
11	+1	+1	-1
19	+1	+1	+1
13	0	0	-2
5	0	0	+2
2	0	+2	0
16	0	+3	0
17	-3	0	0
18	-3	0	0
10	-2	0	0
15	-2	0	0
1	+2	0	0
12	+2	0	0
9	0	0	0
20	0	0	0

[a]coded levels are:	-3	-2	-1	0	+1	+2	+3
calcium, mg dl^{-1}:	5.8	7.8	9.8	11.8	13.8	15.8	-
magnesium, mg dl^{-1}:	-	-	1.0	3.0	5.0	7.0	9.0
protein, g dl^{-1}:	-	3.8	4.8	5.8	6.8	7.8	-

protein levels, respectively. The parameter estimates $\underline{b}_1$, $\underline{b}_2$, and $\underline{b}_3$ are thus measures of the sensitivity with respect to calcium, magnesium, and protein levels, respectively.

During the optimization stage, the following objective function was maximized:

$$F = \underline{b}_1 - |\underline{b}_2| - |\underline{b}_3| \tag{2}$$

Boundary conditions were specified for $|\underline{b}_2|$ and $|\underline{b}_3|$ to keep them less than 10% of the value of $\underline{b}_1$. A boundary condition was also specified for the absorbance of the reagent blank (see Objectives). No boundary was placed on the linearity.

Optimization

The six controllable factors associated with the measurement process itself (HCL-B, 8HQ-B, HCL-A, 8HQ-A, CPC, and DEA) were varied according to the rules of a variable size sequential simplex algorithm (13, 14). The simplex technique has been used previously for laboratory optimizations and is described in the literature (13-20).

After evaluating a predetermined number of vertexes (*i.e.*, after carrying out 25 different treatment combinations of factors associated with the measurement process itself, each time running the tray of 20 samples, fitting Equation 1, and evaluating Equation 2), the objective function had increased by approximately 67% over the worst vertex in the initial simplex. Most of this increase came about because the calcium sensitivity was increased by approximately the same percentage. Magnesium sensitivity increased slightly and protein sensitivity decreased slightly; both were kept well within the established boundary conditions. The baseline response was kept low. Linearity suffered somewhat.

Mapping

After presumably optimum conditions were established by the simplex technique, a Box-Behnken design (21) was evaluated in the region of the suspected optimum to establish tolerances for the six controllable factors and their interactions. The design is a highly fractional, three-level factorial design and is described in the literature (21); it contained 54 treatment combinations, each of which required measuring the 20 serum samples and fitting Equation 1.

The results of the mapping study were used to fit six equations of the form

$$R = \alpha_0 + \sum_{i=1}^{6} \alpha_i y_i + \sum_{i=1}^{6} \sum_{j=1}^{6} \alpha_{ij} y_i y_j \quad (3)$$

where R is one of the six responses considered (objective function, calcium sensitivity, magnesium sensitivity, protein sensitivity, absorbance of reagent blank, or linearity) and the y_i's correspond to the six controllable factors HCL-B, 8HQ-B, HCL-A, 8HQ-A, CPC, and DEA.

Discussion

The data acquired during the mapping study contains information relating six responses to six factors. The results thus contain information about 36 single-factor effects and 90 two-factor interactions. Only a small subset of this information will be discussed in this paper, the apparent effects of the two factors HCL-B and 8HQ-B on the four responses F, b_1, b_2, and b_3.

Figure 5 is a contour plot of the objective function calculated using Equation 3 *vs.* HCL-B and 8HQ-B,

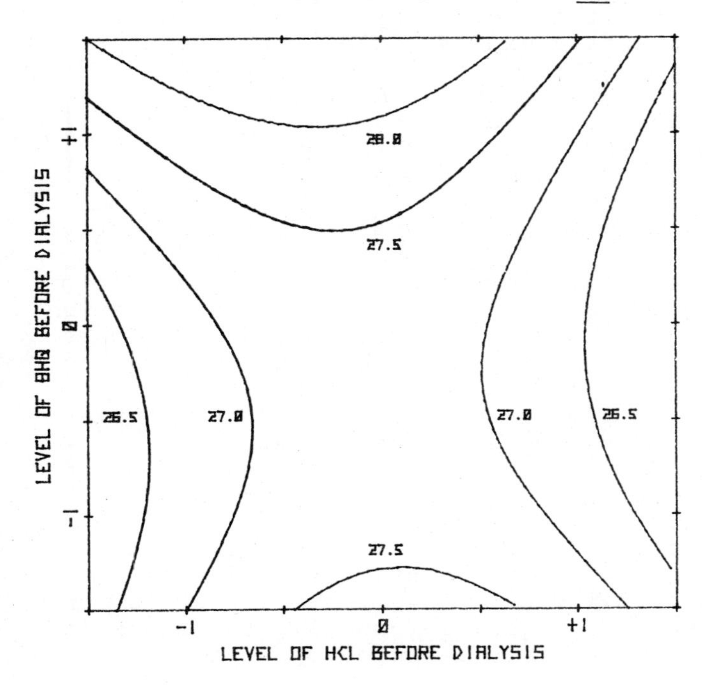

Figure 5. Objective function contours vs. HCL-B and 8HQ-B. Numbers are F × 1000 (see Equation 2).

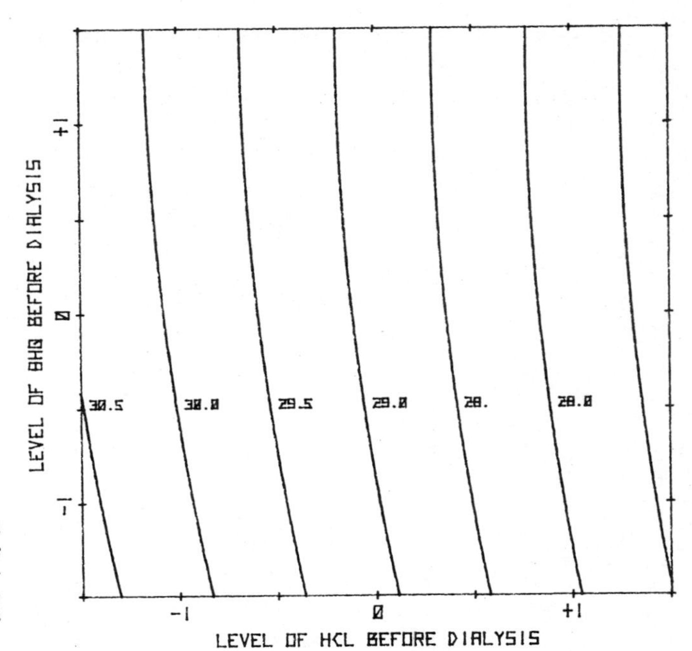

Figure 6. Calcium sensitivity contours vs. HCL-B and 8HQ-B. Numbers are b_1 $\times$ *1000* (see *Equation 1*).

all other factors set at their zero level. The contour is indicative of a saddle region with the stationary point located near the center of the Box-Behnken design. At the zero level of 8HQ-B, the objective function goes through a maximum as the level of HCL-B is increased. At the zero level of HCL-B, the objective function goes through a minimum as the level of 8HQ-B is increased. These results from the mapping study indicate that the value of the objective function could be increased if the amount of 8HQ-B were increased. Recalling that the objective function F is a function of three other responses (see Equation 2), let us examine these three contributing responses in some detail, remembering that the contours are not "exact" as shown in the figures but rather are "fuzzy" because of experimental uncertainties and the possible lack of fit of the model to the data.

Figure 6 shows that HCL-B has a strong effect on the calcium sensitivity; as the concentration of hydrochloric acid before dialysis is increased, the calcium response decreases. The factor 8HQ-B has little effect on the calcium sensitivity; 8-hydroxyquinoline does not bind calcium effectively at low pH.

Figure 7 suggests that, in general, the magnesium interference is not greatly affected by either HCL-B

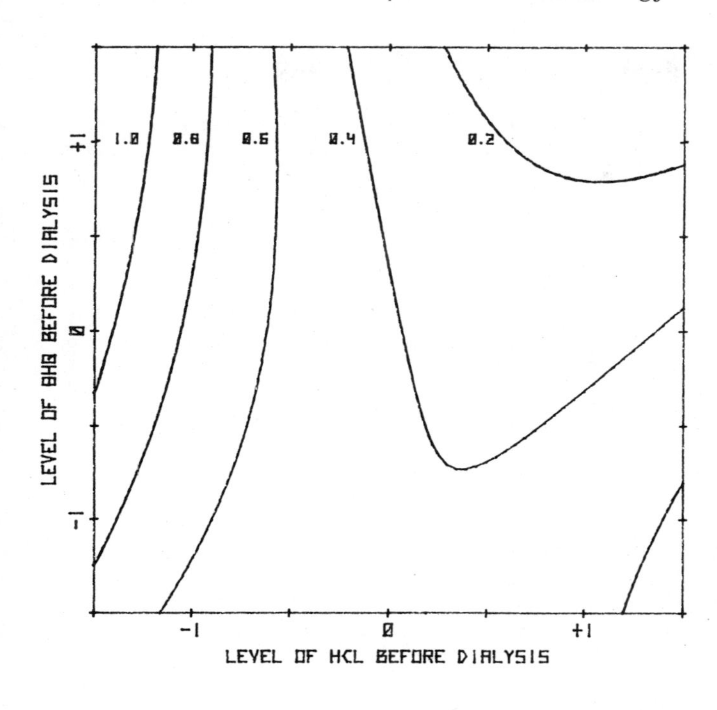

Figure 7. Magnesium sensitivity contours vs. HCL-B and 8HQ-B. Numbers are $b_2 \times$ *1000* (see *Equation 1*).

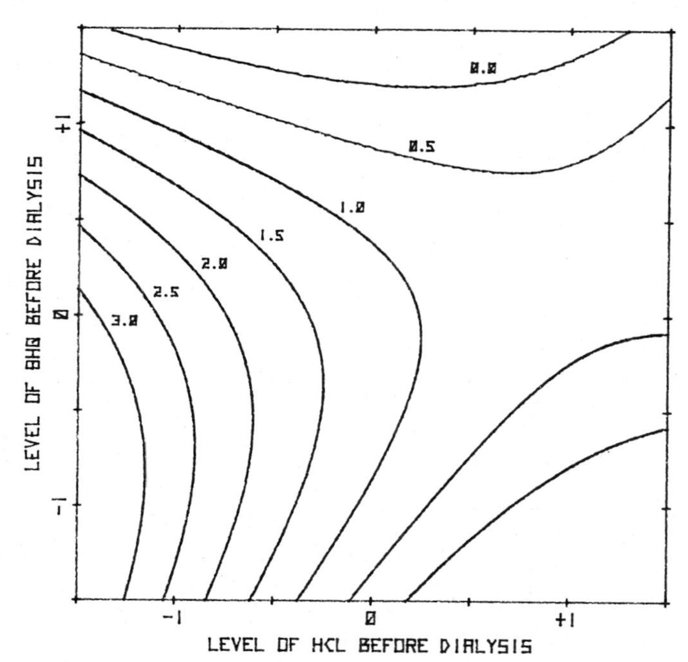

Figure 8. Protein sensitivity contours vs. HCL-B and 8HQ-B. Numbers are $b_3 \times$ *1000* (see *Equation 1*).

or 8HQ-B. Thus, although the absolute value of the magnesium sensitivity is subtracted from the calcium sensitivity in forming the objective function, in this region of the response surface its degradation of the objective function is small. The magnesium sensitivity is well within the boundaries established for it. It is predicted from this figure that the magnesium sensitivity would be near zero at high HCL-B and high 8HQ-B.

Figure 8 shows that HCL-B apparently has a large effect on the protein sensitivity while 8HQ-B has a lesser effect. The large values of protein sensitivity at low HCL-B are subtracted from the calcium sensitivity in this region (see Equation 2 and Figure 6) and causes the objective function to "fold down" at low HCL-B.

Conclusion

Although only a small fraction of the total information obtained in this study has been discussed here, it is evident that classical experimental designs used in conjunction with optimization techniques and computerized automation offer an efficient means of more completely developing and improving chemical measurement processes.

Acknowledgments

The authors thank Ad S. Olansky and Lloyd R. Parker, Jr., for construction of the instrumentation, computer programming, and evaluation of experimental results. This work was supported in part by a grant from Technicon Instruments Corporation and by research Grant MPS74-23157 from the National Science Foundation.

Literature Cited

1. Gottschalk, G., and Marr, I. L., Talanta (1973), 20, 811.
2. Campbell, N. R., "Foundations of Science," 267, Dover, New York, 1957.
3. Mandel, J., "The Statistical Analysis of Experimental Data," 11, Interscience, New York, 1964.
4. Youden, W. J., Materials Research and Standards (1961), 1, 862.
5. Box, G. E. P., Analyst (1952), 77,879.
6. Wilson, A. L., Talanta (1971), 17, 21.
7. Skeggs, L. T., Jr., Am. J. Clin. Pathol. (1957), 28, 311.

8. Snyder, L., Levine, J., Stoy, R., and Conetta, A., Anal. Chem. (1976), 48, 942A.
9. Technicon Instruments Corporation, Method No. SE4-0003FJ4, September, 1974.
10. Anderegg, G., Flaschka, H., Sallmann, R., and Schwarzenbach, G., Helv. Chim. Acta (1954), 37, 113.
11. Lott, J. A., and Herman, T. S., Clin. Chem. (1971), 17, 614.
12. Box, G. E. P., and Wilson, K. B., J. Roy. Statist. Soc., B (1951), 13, 1.
13. Spendley, W., Hext, G. R., and Himsworth, F. R., Technometrics (1962), 4, 441.
14. Nelder, J. A., and Mead, R., Computer J. (1965), 7, 308.
15. Ernst, R. R., Rev. Sci. Instrum. (1968), 39, 988.
16. Long, D. E., Anal. Chim. Acta (1969), 46, 193.
17. Deming, S. N., and Morgan, S. L., Anal. Chem. (1973), 45, 278A.
18. Morgan, S. L., and Deming, S. N., Anal. Chem. (1974), 46, 1170.
19. Yarbro, L. A., and Deming, S. N., Anal. Chim. Acta (1974), 73, 391.
20. Parker, L. R., Jr., Morgan, S. L., and Deming, S. N., Appl. Spectrosc. (1974), 29, 429.
21. Box, G. E. P., and Behnken, D. W., Technometrics (1960), 2, 455.

2

ARTHUR and Experimental Data Analysis: The Heuristic Use of a Polyalgorithm

A. M. HARPER, D. L. DUEWER,* and B. R. KOWALSKI
Laboratory for Chemometrics, Department of Chemistry,
University of Washington, Seattle, WA 98195

JAMES L. FASCHING
Department of Chemistry, University of Rhode Island, Kingston, RI 02881

Most non-routine data analysis in chemstry is designed to aid the formulation and/or evaluation of some model or hypothesis of the intrinsic data structure. The more detailed the model of the data's structure, that is, the more complete the analyst's understanding of the data, the more facile the selection of appropriate algorithms for the data analysis. Conversely, where very little is known of the data's structure it is difficult to make *a priori* selection or evaluation of analysis methodologies.

ARTHUR (1,2), a system of data manipulation, pattern recognition and robust statistical algorithms, is designed as a tool for the analyst in applications where the data's structure is not well understood. The algorithms included in the system are those which our laboratory and other members of the Chemometrics Society have found useful in the analysis of a number of quite different chemical and biological data sets. Recently implemented provisions for the inclusion of measurement uncertainties in the mathematical methods (3) enable the determination of which aspects of the data structure are truly inherent to the data. Descriptions of these algorithms can be found in the appendix of this chapter. It should be noted that ARTHUR is meant to be complementary to and not in competition with such primarily statistical systems as SPSS (4) and BMD (5).

The primary utility of ARTHUR being in the formulation and evaluation of models for incompletely-understood data sets, it is not possible to specify given algorithms or sequences of algorithms which are "best". However, in the course of much data analysis (both fruitful and frustrating) some "rules of thumb" or heuristic procedures have been formulated. Following an introduction to the "ARTHURian" terminology of data analysis and pattern recognition, and a description of the inclusion of measurement uncertainties in pattern recognition methods, the heuristic techniques the developers and users of ARTHUR have found most generally useful will be described.

*Present address: Department of Chemistry, University of Arizona, Tucson, Arizona 85721.

Definitions

The following terms and definitions have proved useful in describing the types of data analysis algorithms available in ARTHUR and in describing the data to be analyzed.

Classification analysis. The data are known to be composed of specified groupings or categories. The goals of such analysis are the identification of what parameters (if any) qualitatively distinguish the known groupings and (if possible) the selection of a classification rule for identifying the known groups.

Continuous property analysis. The data are known to represent a continuous range of responses towards some given property(ies). The goals of such analysis are the identification of what parameters (if any) are functionally related to the property and (if possible) the selection of a rule which quantitatively predicts that property.

Unsupervised analysis. The data are not known to have any systematic characteristics. The goal of such analysis is the discovery of what systematic behavior the data exhibit (if any exists). Study of the regularities among objects is generally referred to as cluster analysis; study of the regularities among measurements is generally referred to as factor analysis.

Object. A compound, sample, individual or other entity for which a list of characterizing parameters is present in the data base.

Measurement. An experimentally available parameter (independent variable) used to characterize the objects.

Feature. Any transformation of one or more measurements used to characterize the objects. When referring to a parameter which can be either a measurement or a feature, the term "measurement/feature" is used.

Data vector. The complete list of measurement/feature values used to characterize a particular object. (The older chemical pattern recognition literature, including that of the Laboratory for Chemometrics, refers to this as a "pattern". Considerable semantic confusion over "patterns of patterns" forced the change to the term, "data vector".)

Category. One of the groups of objects studied in the classification analysis algorithms. Categories which are entirely independent of one another, such as the labeling of white bond papers by their manufacturer, are referred to as discrete categories. Categories which have some dependence upon one another, such as "low, middle and high," are referred to as continuous categories.

Property. A quantitative parameter characteristic of the objects for which a functional representation is desired (dependent variable).

Training set. The list of data vectors used to generate classification or prediction rules.

Evaluation set. The list of data vectors used to evaluate the performance of classification or prediction rules.

Test set. The list of data vectors, in classification or prediction analysis, for which the true category or property value is not known. The Evaluation and Test sets are functionally one and the same. The Evaluation set is a "let's pretend" Test set.

Uncertainty. The error associated with an analytical measurement. The uncertainty is assumed to include all sources of errors such as sampling, instrumental, chemical, etc.

It should be recognized that these definitions are not particularly rigid or mutually exclusive. A continuous property can certainly be segmented into the low resolution categories "too low" and "high enough". The parameter considered as a property in one phase of analysis may well be a measurement in another. The Training and Evaluation set definitions may be switched. It may even be desired to switch the definition of object and feature. If the data are considered as a matrix (objects as rows and features as columns), the switch is equivalent to the transposition of the matrix. And it is certainly good practice, no matter what the specific nature of the data analysis problem, to make at least cursory unsupervised data analysis, if nothing more than to give a rough screen for some gross, unsuspected structure in the data.

Pattern Recognition: New Techniques that Utilize Analytical Error

The general problem that is amenable to solution by the techniques available in ARTHUR is the analysis of patterns in an n-dimensional space. In the past, applications utilizing pattern recognition have not taken into account the errors in the measurements because the mathematical methods currently available make no provision for their inclusion. However, in most chemical data the inadvertent assignment of zero measurement error which results is clearly an unrealistic assumption. This problem has been investigated by Fasching, Duewer and Kowalski (3). As a result of this study, several algorithms in ARTHUR have been modified to include the uncertainties in the calculations.

Current pattern recognition techniques treat measurements as dimensions in an n-dimensional space. If, for each member of a collection of objects (samples), n measurements are known, the samples are represented as points in the space formed by the measurements. Therefore the value of a given measurement for a particular object serves to exactly position the point representing the object along a coordinate measurement axis in the n-space. Figure 1 depicts the configuration of the data vectors from two samples in a three-dimensional space. The set of all such vectors defines the data matrix.

In analytical applications, where the uncertainties in the measurement are either known or can be estimated, there exists a matrix of uncertainties corresponding to and of the same dimensions as the data matrix. Mathematical operations that transform the data matrix also change the uncertainty matrix to a transformed uncertainty matrix. Each element of the uncertainty matrix reflects the exactness (in units of $\pm$ one standard deviation) to which the corresponding element of the data matrix is known. Therefore, each measurement in the data matrix is now treated as a mean value with a probability distribution defined by its error. The effect of the inclusion of uncertainties on the vectors in Figure 1 is illustrated in Figure 2. The analytical uncertainties reflect the fact that a data vector is not, in reality, a point in the measurement space, but is the most probable value in a region of probability in this space. If the area of the elipsoid in this example is defined at a 50% probability level of the standard deviation of each measurement, then another set of measurements made on a sample would have an equal probability of lying outside the elipsoid as within it. This model is more reasonable for most chemical problems.

At present, ARTHUR has been modified to include the analytical error in representative methods of preprocessing, display, supervised learning and unsupervised learning. A full description of these modifications can be found in reference 3. The current methods deal only with symmetric uncertainties. A nonmetric (unsymmetrical) distance is defined; however, classification and clustering routines utilizing distance have not, as yet, been similarly modified to make use of this type of distance matrix.

Since the uncertainty matrix contains information about the error associated with each measurement, it can be incorporated into the preprocessing of the data matrix. The more realistic features generated can be utilized in all reported methods of pattern recognition, thus eliminating the need to change each analysis method. The scaling algorithms (SCALE)* in ARTHUR have been modified to include uncertainties. An error-weighted mean and variance are utilized in place of the feature mean and variance in these calculations. The new mean of the jth feature in the data is defined as:

$$\bar{x}_j = \frac{\sum_{i=1}^{m} x_{i,j}/u^2_{i,j}}{\sum_{i=1}^{m} 1/u^2_{i,j}}$$

*Methods (names in capital letters) are described in Appendix.

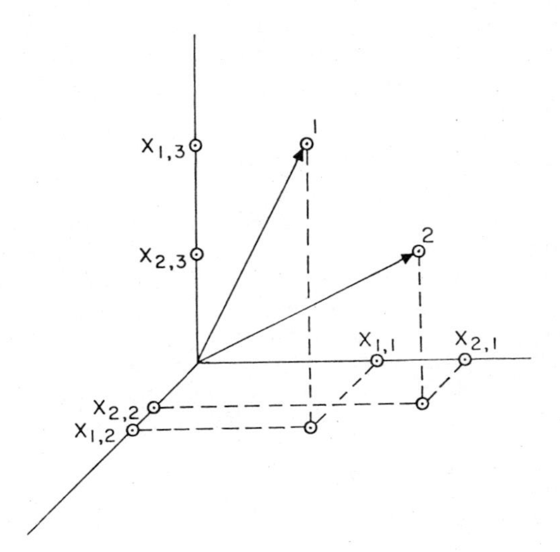

Figure 1. Data vectors in three dimensions

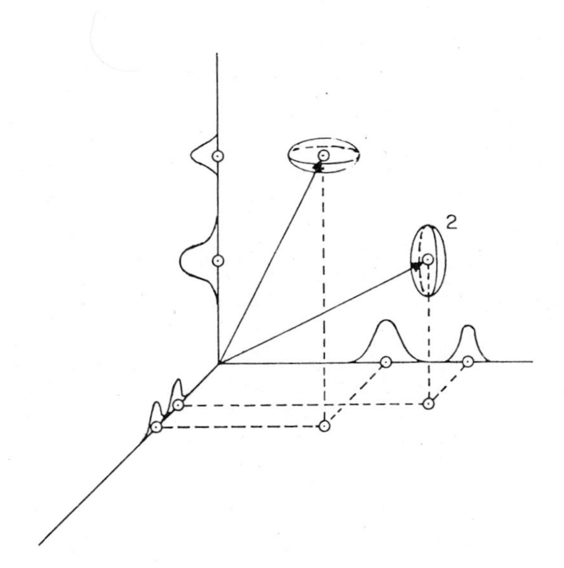

Figure 2. Elipsoids in three dimensions

where the $u_{i,j}$'s are the entries in the uncertainty matrix corresponding to the data matrix measurement $x_{i,j}$, and the sum is over the training set data vectors. Modification of the available distance metrics have also been made along with the addition of new distance calculations based on measurement uncertainties. The algorithms for these can be found in the appendix (DISTANCE) to this chapter. The modified city-block distance and the modified Mahalanobis distance are now weighted by a function of the measurement errors associated with the features going into the calculation. A new metric, the gaussian overlap-integral distance, greatly emphasizes the features that have a small distribution with respect to their measurement size and related uncertainties. A maximum distance of one is assigned to features that differ greatly from each other or have very small uncertainties. Another new distance calculation, the gaussian feature-space distance, calculates a distance value that is proportional to the probability that a feature in the ith data vector belongs to the same population as the corresponding feature of the jth data vector. These are summed over all the feature space to give the intersample distance. The calculation is nonmetric and the distance matrix is unsymmetrical.

The uncertainty matrix has also been incorporated in the Karhunen-Loève transform. The modified technique transforms the uncertianties into a new uncertainty matrix along with the sample matrix. The assumption is made that the same degree of correlation applies to the uncertainty matrix as is used in the transformation of the sample matrix.

Introduction to Data Analysis Using ARTHUR

Different collections of objects may have quite different data sturctures varying from a random scatter to well defined clusters or curvilinear shapes. Since each algorithm affects data reduction according to the criterion upon which it is based, a thorough understanding of the inherent assumptions imposed upon the data structure in the formulation of a technique and the limitations that may result can provide information helpful in arriving at an understanding of the underlying structure of the data when the methods are applied in combination.

Suppose, for example, the n-dimensional structure of two categories of objects we wish to separate by pattern recognition classification techniques corresponds to the one dimensional problem depicted in Figure 3, where the shaded portions of the figure correspond to category 1 and the unshaded portions to category 2. Whereas in one dimension the solution to the problem is obvious, in n-dimensions the bimodality may not readily reveal itself. If PLANE or SIMCA were applied to these data, the results might lead one to believe that the categories cannot be distinguished since the data are neither linearly separable nor continuous. However, KNN would encounter no problems since the objects in the near vicinity of a given point tend to be of its class.

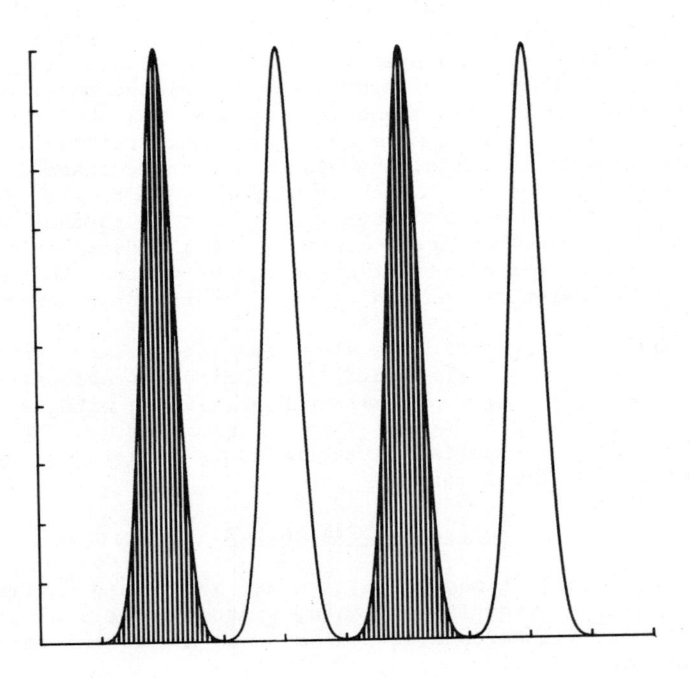

Figure 3. Bimodal distribution

Bayesian classification (as long as no a priori distribution is assumed) would also produce good results. (Note that plots of the data might expose this distribution in a less ambiguous form. Consequently, this example is meant only as an illustration of the effects of the methods on an easy to understand distribution)

Unfortunately, the solution to a real problem does not, in general, tend to be as straightforward and may require a great deal of interaction and guidance from the analyst aided by preprocessors, display methods, and statistics. For this reason, the capabilities of ARTHUR for displaying the data are quite well developed when combined with the ingenuity of the analyst as will be seen in a later section of this chapter. On the other hand, since preprocessing refers to any method that translates, rotates, or in any way transforms the data, such infinite diverse possibilities arise that were we to include only those methods that we and others have found useful, they would dominate the code. Therefore, the set of preprocessing tools available in ARTHUR is aimed mainly toward normalization, feature weighting, and dimensionality reduction. In addition, ratios of features can be added to the feature list in TUNE and individual features can be transformed or combined in CHFEATURE. Since the methods chosen to preprocess the data can ultimately determine success or failure in the solution of a problem by pattern recognition methods and/or the cost of the analysis, methods not available in ARTHUR should not be neglected. An example of this is the utilization of the Fourier, Hadamard and autocorrelation functions for the transformation of spectral data (6,7).

Two assumptions made throughout any supervised pattern recognition technique are that the features used contain information useful to the solution of the problem and that, even when this is known to be the case, the data can be transformed into a representation amenable to the algorithms employed. When this is not the case it may become necessary to either change the form of the question being asked about the existing data or redesign the experiment from which the measurements are obtained. Hopefully, the information gained through prior analysis will serve to guide the analyst in this endeavor.

We have discovered that techniques originally designed for unsupervised learning applications are powerful tools in the early stage of all data analysis problems. These methods have seen little application in chemistry. Since the goal of these methods is the determination of the existence of inherent data structures within a larger data structure, neither training nor classification is attempted. TREE and HIER are two unsupervised learning methods which are based on the similarity of objects as defined by their distances in the feature space. Factor analysis can also be utilized in this mode.

We have found the following procedures to be generally useful in data-model formulation. Note that many of the algorithms are multi-purpose; one application of the algorithm can, however, be made to serve more than one purpose with careful planning (and some luck). It must also be noted that certain algorithms, in particular the measurement-by-measurement plots (VARVAR) and hierarchical clustering (DISTANCE-HIER), can become very expensive of computer and/or analyst resources with large data bases.

I. Input and validate the data. Applies to ALL data bases.
  A. Get the data from coded-file ("card image") to binary format (INPUT).
  B. Check for missing data (INFILL).
  C. Check for constant and/or exactly redundant measurements (INDUMP).
  D. Check for invalid or suspect data by looking for outliers and unique behavior (UTILITY; SCALE; VARVAR; principal component analysis: SCALE, KAPRIN, KAVECTOR, KATRAN, VARVAR; Cluster analysis: DISTANCE/HIER, or DISTANCE/TREE).
  E. Correct any problems identified in above procedures and try again.

II. Preliminary examination of measurements. Applies to MOST data bases.
  A. Get the univariate measurement statistics and distributions (SCALE and BASET-BAHIST).
  B. Get the measurement-by-measurement plots (VARVAR).
  C. Modify or remove any measurements identified as "intractible" in above procedures, particularly necessary with qualitative or quantified measurements (CHFEATURE).
  D. Scale all measurements to uniform mean and or variance (SCALE).
  E. Examine the intermeasurement correlations (CORREL; HIER, -TREE, -VARVAR; Principal component analysis: KAPRIN-KATRAN-VARVAR, -KAVECTOR).
  F. Determination of the intrinsic dimensionality of the data (KAPRIN-KAVECTOR, -KAMALIN, -KACROSS).
  G. Produce a reduced, orthogonal set of features (KAPRIN-KATRAN, -KAVARI-KAORTH-KATRAN).

III. Category Classification.
  A. Examine the univariate Fisher and variance weights (WEIGHT).
  B. Modify or eliminate measurement features with trivial univariate classification utility (CHFEATURE; WEIGHT-CHFEATURE, -GRAB; SELECT).
  C. Linear Discriminant classifications (PLANE, MULTI, LESLT, LEDISC, REGRESS).
  D. Projection of data onto discriminant axes (LESLT-VARVAR, -SCALE-KAPRIN-KATRAN-VARVAR).
  E. k-nearest neighbor analysis (DISTANCE-KNN).
  F. Category modeling with principal component descriptions

(SIJACOBI-, SIPRINCO-SICLASS; SICSVA).

G. Bayesian classification (BASET-BACLASS, -BAGAUS-BACLASS, -BASPLINE-BACLASS).

H. Modify or delete measurement features as indicated during above analysis and try again (CHFEATURE, TUNE, TUMED).

I. Redefine or delete data vectors as indicated during above analysis and try again (CHDATA, CHCATEGORY, CHSPLIT, CHJOIN).

J. COMPARE THE RESULTS OF THE VARIOUS METHODS! The differences between the various classification algorithms' results may indicate what data-structure assumptions are valid, and may lead to significant progress in the understanding of the intrinsic structure.

K. Using measurement/features selected to give optimum classification, perform cluster analysis (DISTANCE-HIER, -TREE, -NLM - VARVAR; KAPRIN-KATRAN-VARVAR).

IV. Property prediction.

A. Examine the univariate correlations-to-property (WEIGHT).

B. Examine all property-to-measurement plots (VARVAR).

C. Modify or eliminate measurement features with trivial univariate prediction utility (CHFEATURE; WEIGHT-CHFEATURE, -GRAB; SELECT).

D. Multilinear regression and measurement selection (LEAST, STEP, REGRESS, DISTANCE-LEPIECE).

E. Nearest neighbor averaging (DISTANCE-PNN).

F. Modify or eliminate measurement features as indicated during above analysis and try again (CHFEATURE, TUNE TUMED).

G. Redefine or delete data vectors as indicated during above analysis and try again (CHDATA).

V. Unsupervised clustering (without prior measurement/feature selection).

A. Cluster analysis using the orthogonal; reduced feature set of II(G) (DISTANCE-HIER, -TREE, -NLM-VARVAR).

B. Postulate some data structure.

C. Perform the appropriate analysis for the hypothesized structure.

D. Optimize the measurement/features for the postulated data structure (CHFEATURE, TUNE, TUMED, SCALE).

E. Redo the cluster analysis with the optimized measurement/features.

F. Be cautious about leaping to conclusions.

VI. Estimate the degree of confidence you can place upon the analysis. Applies to MOST data bases.

A. Many of the classification, prediction and factor analysis algorithms provide statistical probability estimates as part of their output. These estimates assume the data to be an adequate representation of the parent population and to be multivariate normal. We have found these parameters to give a much better indication of how poor training set performace was rather than how good test set performance will be.

B. The performance of classification, prediction and some factor analysis algorithms can be qualitatively estimated from the Evaluation Set results. The degree that these results can be used as an indication of future performance is limited by how adequately the parent population is represented by the Evaluation set.

C. If the number of objects in the Training and Evaluation sets is sufficiently large, jackknifing (8) can be used to estimate classification, prediction and factor analysis reliability (CHSUBS, CHDATA).

D. If the experimental data uncertainties are known or can be estimated, a Monte Carlo technique we refer to as "Uncertainty Perturbation" (9) may be used with all analysis algorithms (CHRANDOM, TURANDOM). Many algorithms make explicit use of experimental uncertainties, if available, as well.

Application of Display Techniques to Obsidian Data.

To illustrate the utility of some of the algorithms suggested in the previous outline for examination of a data set, the "much used" ARCH obsidian data described in references 10 and 11 were examined by the sequence of routines shown in Table 1. This data set, because of its simple structure, accompanies the send-out version of ARTHUR (1,2). ARCH is separable by all classification routines in ARTHUR and is used in our laboratory as a debugging and method testing aid. Table 2 gives some of the statistics for the obsidian data (INPUT-SCALE). It consists of elemental concentrations of ten metals in 63 obsidian samples from 4 sites near San Francisco, obtained by X-ray fluorescence.

Histogram plots (BASET-BAHIST) provide a convenient method for rapid scanning of the feature distributions. Both the individual category and the summation-over-the-data histograms are output from BAHIST. Figure 4 is the summation plot for Barium. There appear to be three modes in the overall distribution. Inspection of the individual-category histograms for this element revealed that within each category the distribution is unimodal with categories two, three and one and four comprising the three modes of the summation plot. In view of this, it is not surprising that this feature receives a high weighting by both the Fisher and variance criteria (SCALE-WEIGHT) as shown in Table 3. Figure 5 displays in dendrogram form the Q-mode clustering of the interfeature correlations (SCALE-KADISP-HIER) of ARCH. In this plot the shorter the "branch" between two features, the higher the correlation. Yttrium shows no correlation with the rest of the features.

Table 1

Number of training vectors	63
Number of test vectors	12
Number of features	10
Ratio of training vectors to features	6.3
Number of categories	4
Average ratio of vectors to features for each category	1.58

Feature List

Feature		Mean	Standard Deviation	Skewness	Kurtosis
1.	Fe	120.9	232.3	.49	1.62
2.	Ti	26.02	111.8	-.04	1.37
3.	Ba	42.48	16.70	-.86	2.77
4.	Ca	680.8	278.6	-.42	1.32
5.	K	392.8	54.51	.05	2.35
6.	Mn	46.52	14.41	.53	2.03
7.	Rb	107.5	16.84	.25	2.42
8.	Sr	31.59	17.36	.32	1.82
9.	Y	56.02	6.987	-.26	2.81
10.	Zr	15.68	43.73	-.82	3.11

Table 2

INPUT	Input data.
SCALE	Autoscale the data.
BASET	Create frequency histograms for each measurement.
BAHIST	Plot the frequency histograms.
WEIGHT	On scaled data, get the ordered Fisher and variance weights.
VARVAR	Plot the Fisher weighted data.
VARVAR	Plot the variance weighted data.
CORREL	Get the lower diagonal correlation matrix.
KADISP	On scaled data, create a distance matrix from the feature correlations.
HIER	Q-mode clustering of feature correlations.
KAPRIN	On scaled data, get the ordered eigenvectors and eigenvalues.
KATRAN	Transform the data by its eigenvectors.
VARVAR	Plot the transformed data. This is the Karhunen-Loève projection.
TUNE	On unscaled data get the n(n-1) ratios of the measurements.
CHJOIN	Form a new data file of n^2 measurements by merger of input and ratio data.
SCALE	Autoscale the n^2 measurements.
SELECT	Select the n most "important" of the n^2 measurements by the total Fisher weight.
VARVAR	Plot the selected features.
ENDIT	Program termination.

Table 3

Feature	Weight: Variance	Weight: Fisher
1. Fe	8.414	1.202
2. Ti	9.005	2.342
3. Ba	37.01	6.648
4. Ca	8.830	2.519
5. K	4.957	.3736
6. Mn	4.792	.3442
7. Rb	2.680	.1685
8. Sr	6.305	.5323
9. Y	1.016	.0011
10. Zr	11.11	1.120

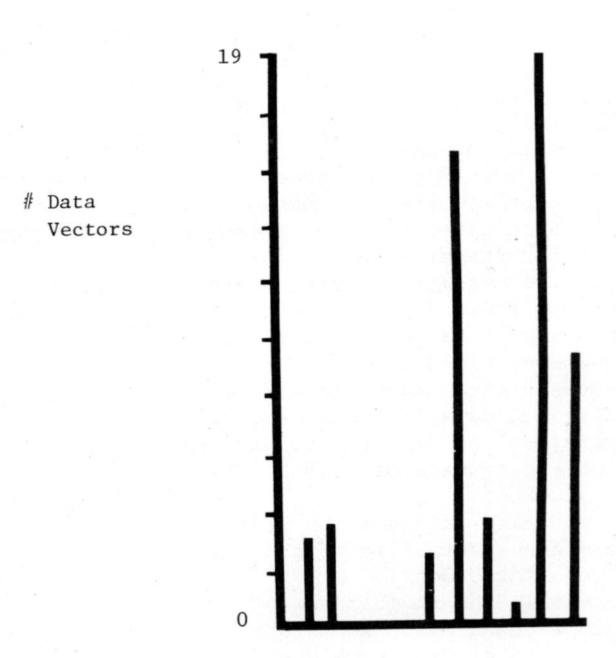

Figure 4. Summation histogram over all categories for barium

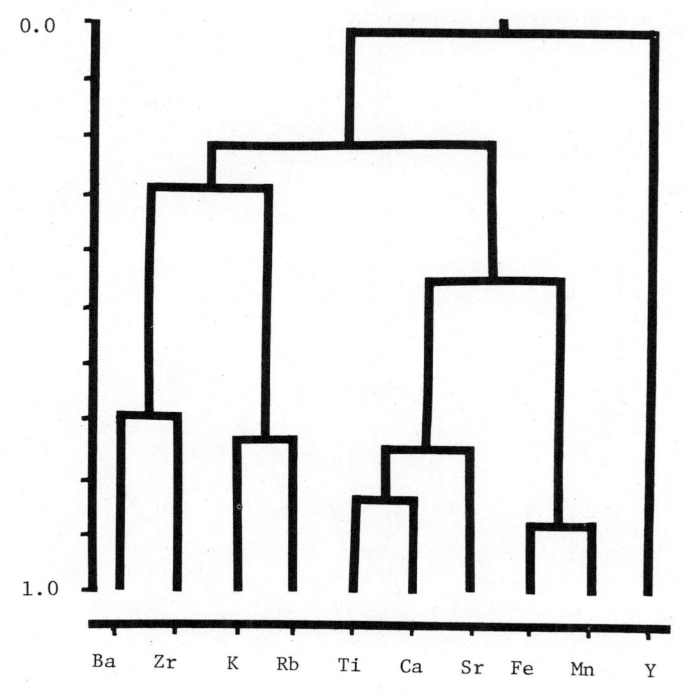

Figure 5. Hierarchical dendrogram of feature correlations

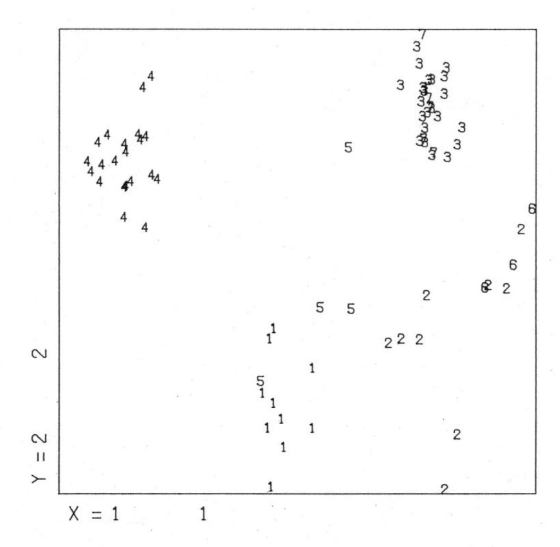

Figure 6. Karhunen–Loève projection (training set categories are denoted by numerical values 1–4 and test vectors by numbers greater than 4)

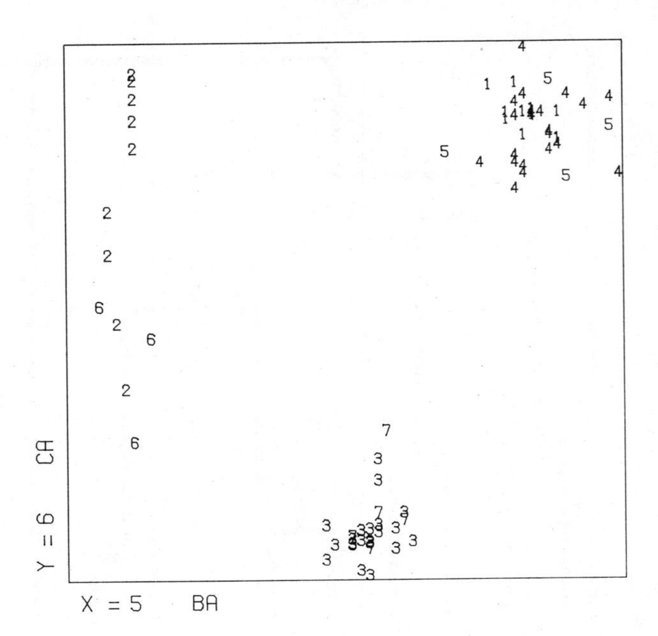

Figure 7. Fisher weighted features

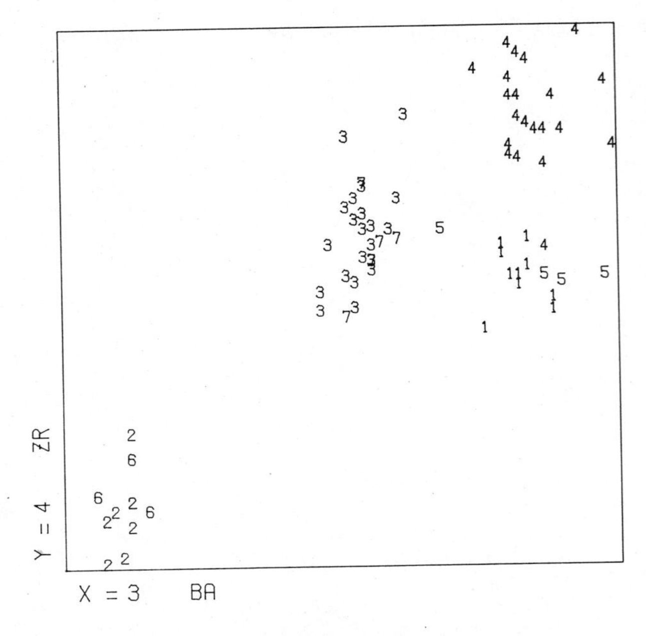

Figure 8. Variance weighted features

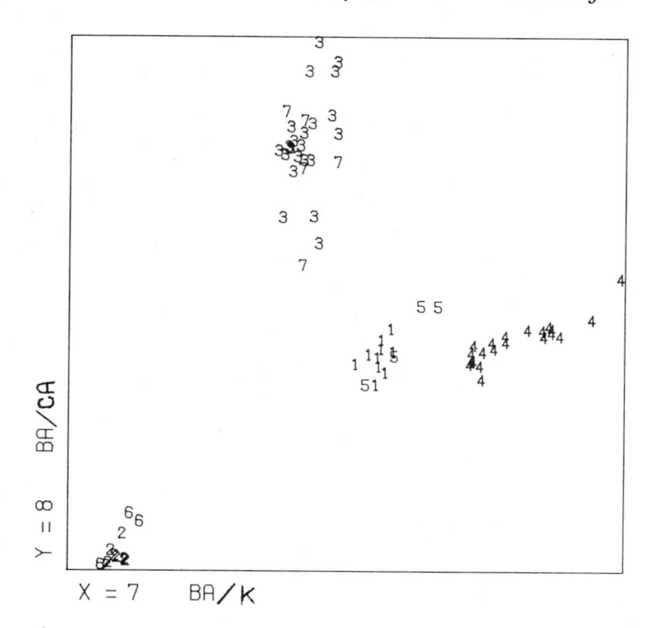

Figure 9. First two features selected from addition of feature ratios to original data

and was shown in Table 3 to have very low utility for category separation. The effect of this in Factor Analysis methods has been discussed by Duewer (9).

The first two features formed from the Karhunen-Loève transform of the data (SCALE-KAPRIN-KATRAN-VARVAR) are plotted in Figure 6. These features are representative of 74.5% of the total variance in the data. The plot shows the individual categories as forming clusters in this space. However, it should be kept in mind that this display technique has as its goal preservation of as much information (in a variance sense) in as few features as possible. Therefore, the features formed are ordered by their overall contribution to the total variance and not by their utility in separation.

Figures 7 and 8 are plots of the first two features of the Fisher and variance weighted data, respectively, of Table 3. WEIGHT, when used as a variable reduction technique, should be approached with caution. The same or more information can be retained by the orthogonal selection offered in SELECT in fewer features. Also, we have discovered that, for data sets of a large number of categories, easy separability between a few of the categories can result in the highly weighted features all describing the same category separations in the data. Therefore, for this type of data, it is a good practice to examine the individual category-pair weight contributions to the total weight. For example, neither of the Fisher weighted features plotted in

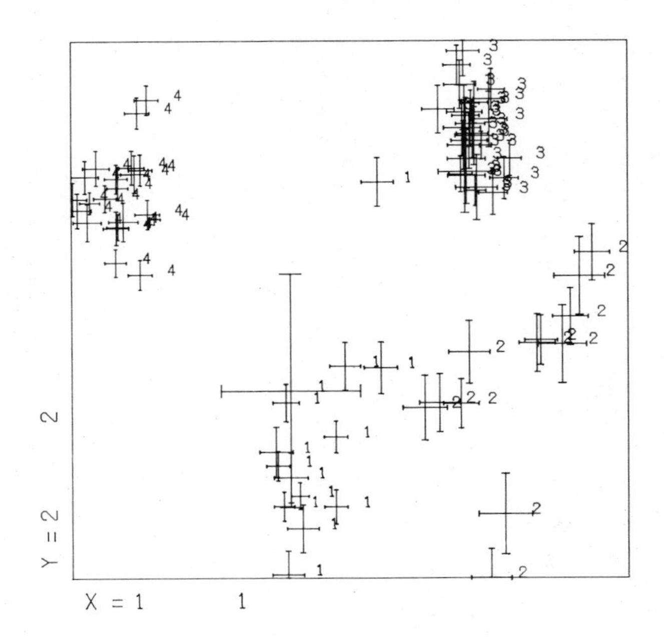

Figure 10. Karhunen–Loève projection of data with uncertainties

Figure 7 describe the separation of Category 1 from Category 4.

Figure 9 shows the first two SELECT features (TUNE, CHJOIN, SCALE, SELECT, VARVAR) chosen from a feature list containing the original measurements plus all ratios of the measurements. The weighting used as a basis for the selection was the total Fisher weight. Although Potassium previously was indicated to have less utility for separation than six of the other metals, the ratio BA/K has a higher weighting than that of Barium. Comparison of this plot with Figure 5 reveals that the addition of the Potassium factor has eliminated the overlap of Categories 1 and 4.

Figure 10 depicts the Karhunen-Loève projection of the same data with the uncertainties added. Measured uncertainties were available for ninety percent of the data and the rest were estimated by one of the authors (JLF). The limits of one standard deviation for the uncertainties of the transformed variables determine the size of the cross-hatch. Samples that display large uncertainties for the new features should be reinvestigated or eliminated from the data.

Conclusion

In March, 1975, our laboratory announced the availability of ARTHUR upon request for a small fee equivalent to the cost of documentation copying, tape-creation computer time, and mailing. It is currently in use at several laboratories around the world.

This version of the program evolved from a need in the scientific community for a versatile, easy to use, pattern recognition system covering the most recent developments in pattern recognition methods.

Earlier versions of the program had been devoted mainly to those methods that fell in the realm of pattern recognition. As we gained experience with the analysis of real problems, ARTHUR expanded to include various multivariate statistical methods. However, the development of a useful pattern recognition program has become a continuous, and at times accelerating, project.

The routines described in this chapter are the major ones currently in use in our laboratory. Since we are concerned with solving chemical problems with data analysis mathematics, ARTHUR has evolved into a more versatile system for studying the n-dimensional data structure of a series of samples.

There are many methods that could be added to the program. This process takes a considerable amount of time because a new method must be programmed, debugged and tested. It must also provide a clear advantage over existing methods in ARTHUR. At the Laboratory for Chemometrics in Seattle, development of the link and support of the system to several different interactive graphics terminals (2) is continuing and several novel applications are in progress.

Alternate methods for handling missing data are being investigated and added to the program. Expansion of pattern recognition techniques to utilize uncertainty is continuing both at the University of Rhode Island and at the University of Washington.

APPENDIX

Bayesian Classification

An approximate Bayes Rule forms the criterion for classification of categorized data vectors in BAYES. The subsystem is made up of seven stand-alone subroutines. Used as a classifier, BAYES first builds frequency histograms for each feature in each category. These may be used in their "raw" form in place of the probability distribution in Bayesian classification or may be smoothed by fitting a Gaussian distribution or a cubic spline function prior to classification.

BACLASS. Classification of data vectors based on frequency histograms is done in BACLASS (12). If P_k is the a priori probability of category "k" and R_k is the risk associated with the misclassification of a pattern in "k," then the probability that a given pattern "i" is a member of "k" is given by:

$$P(X_k|x_i) = \prod_{j=1}^{n} \left[\frac{P_k R_k P(x_{i,j}|X_{j,k})}{\sum_{m=1}^{\ell} P_m R_m P(x_{i,j}|X_{j,m})} \right]$$

where "ℓ" is the number of categories, "n" is the number of features and $P(x_{i,j}|X_{j,k})$ is the value of the probability distribution for the jth feature of category "k" at the value $x_{i,j}$.

An alternative probability calculation provided is:

$$P(X_k|x_i) = \sum_{j=1}^{n} \left[\frac{P_k R_k (P(x_{i,j}|X_{j,k}))}{\sum_{m=1}^{\ell} P_m R_m \ (P(x_{i,j}|X_{j,k}))} \right]^{\alpha}$$

where "α" is user defined. In either case, the data vector is assigned to a class according to the largest value of $P(X_k|x_i)$.

BASET. BASET constructs the raw frequency histograms from the input data file training set vectors (12). Any assumption of feature independence will be true only if it is true for the input data. A histogram is built for each variable in each category. The resolution of the histograms is determined by the total number of data vectors in the smallest category. (By default, it is one-fifth of this number, but may be redefined by the user.) The intensity of a particular resolution element in a histogram is based upon the frequency of feature values within a category that lie within its range. The histograms are stored on a disk file.

BAGAUS and BASPLINE. These two routines allow the user to smooth the histograms built in BASET. Gaussian smoothing is done in BAGAUS, whereas BASOLINE performs a cubic spline fit (13). The parameters obtained from the fits are stored on a file and can be used for classification in BACLASS.

BAHIST. Line printer histograms for each variable of each category are provided by BAHIST (12). These are displayed with normalization by both the category maximum and training data maximum. Summation histograms for each feature over all the data are also printed.

BATEST. BATEST is used to examine each histogram for normality. A theoretical true Gaussian distribution is built from the standard deviation and mean of the input histogram. The histogram is then scanned for the element that represents the maximum deviation from the theoretical distribution. A similarity value between 0 and 1 is calculated using the Kolomogorov-Smirnov test for normality (14). This test is sensitive to any "holes" in the histogram, and should be used along with cross-checking by visual inspection of the histogram.

Intermeasurement Correlations

CORREL calculates all feature-feature and feature-property covariances and correlations. Confidence intervals for each correlation coefficient and estimates of the probability that the data could come from an uncorrelated parent population are also provided. If X is a data matrix of m data vectors, then the covariance between the ith and j^{th} feature in X is:

$$(CV)_{i,j} = \frac{\sum_{k=1}^{m} (x_{i,k}-\bar{x}_i)(x_{j,k}-\bar{x}_j)}{m-1}$$

and the interfeature correlation of features i and j becomes:

$$C_{i,j} = \frac{\sum_{k=1}^{m} (x_{i,k}-\bar{x}_i)(x_{j,k}-\bar{x}_j)}{\left[\sum_{k=1}^{m} (x_{i,k}-\bar{x}_i)^2 \sum_{k=1}^{m} (x_{j,k}-\bar{x}_j)^2\right]^{1/2}}$$

For data whose properties are continuous, the correlation of the individual features to the properties is informative. This is defined as:

$$(CP)_i = \frac{\sum_{k=1}^{m} (x_{i,k}-\bar{x}_i)(p_k-\bar{p})}{\left[\sum_{k=1}^{m} (x_{i,k}-\bar{x}_i)^2 \sum_{k=1}^{m} (p_k-\bar{p})^2\right]^{1/2}}$$

where the sum is for the ith feature over the m data vectors, P_k is the property associated with the kth data vector, and $\bar{p}$ is the average for the property over all of the data.

A confidence interval about the correlation is obtained with the Fisher z-transform (15). If $C_{i,j}$ is the correlation between the ith and j^{th} feature then

$$z = \tanh^{-1} (C_{i,j})$$

is approximately normally distributed with standard deviation

$$\sigma_z = (m-3)^{-1/2}$$

The confidence interval $(CI)_{i,j}$ about the correlation becomes

$$\pm (CI)_{i,j} = \tanh (z \pm t\sigma_z)$$

where "t" is Student's t value.

The computed correlation coefficient can be used to test the hypothesis that the data could have come from a parent population having a zero correlation (16). The marginal distribution density of the correlation coefficient $C_{i,j}$ is:

$$P_{C_{i,j}}(m,C) = \frac{1}{\pi} \frac{\Gamma[1/2(m-1)]}{\Gamma[1/2(m-2)]} (1-C_{i,j}^2)^{(m-4)/2}$$

thus the probability of rejection of the hypothesis that the true correlation is zero when it is true becomes:

$$P_{rej} = 2\int_{|C_{i,j}|}^{1} P_{C_{i,j}}(m,c_{i,j})\, dC_{i,j}$$

Interpattern Distance Matrix Calculations

DISTANCE calculates the interpattern distance matrix and stores it for use in other routines. For data with known uncertainties, several distance metrics are available which take into account the measurement error. The distance metrics for data with no uncertainties include:

Mahalanobis distance of order N: (17)

$$D(N)_{i,j} = \left[\sum_{k=1}^{n} (x_{k,i} - x_{k,j})^N \right]^{1/N}$$

where $x_{\ell,m}$ is the ℓth variable of the mth data vector and n is the total number of features. For N=2 this is the Euclidean distance.

City block distance:

$$D_{i,j} = \sum_{k=1}^{n} |x_{k,i}-x_{k,j}|$$

Ratio distance of O. U. Anders (18):

$$D_{i,j} = \frac{\left[\sum_{k=2}^{n} \sum_{\ell=1}^{k-1} \delta_{i,j,k,\ell}\right]}{[n(n-1)/2]}$$

where $\delta_{i,j,k,\ell} = 0$, when $LL < \dfrac{(x_{k,i}/x_{\ell,i})}{(x_{k,j}/x_{\ell,j})} < UL$

$= 1$, when above is not true.

LL and UL are the lower and upper limits, respectively, for "matching" of the ratios. These limits are selected by the program user.

The distance metrics for data that have uncertainties are (3):

Behren-Fisher: This algorithm is a Mahalanobis distance function of order n, weighted by measurement uncertainties. The distance (d_{ij}) between data vectors i and j is:

$$d_{ij} = \left[\sum_{k=1}^{m} \left[\frac{x_{ik}-x_{jk}}{\alpha+\beta(u_{ik}^2+u_{jk}^2)^{1/2}}\right]^n\right]^{1/n}$$

where "m" is the number of features, α and β are weighting constants (usually equal to one) and $u_{\ell k}$ is the uncertainty associated with the kth feature ($x_{\ell k}$) of the ℓth data vector.

Error weighted city block: The uncertainty-modified city block distance (d_{ij}) is:

$$d_{ij} = \sum_{k=1}^{m} \frac{|x_{ij}-x_{jk}|}{\alpha+\beta(u_{ik}^2+u_{jk}^2)^{1/2}}$$

where m, x_{ik}, u_{ik}, α, and β are the same as for Behren-Fisher.

Gaussian Overlap-Integral Distance: This distance calculation greatly emphasizes the features having a small distribution with respect to their measurement size and uncertainties. If x_{ik} is the value of the ith feature of the kth data vector and u_{ik} is the related uncertainty in x_{ik}, the Gaussian probability function for feature x_{ik} is:

$$P_i(x_{ik},\bar{x}_{ik},u_{ik}) = \frac{1}{u_{ik}\sqrt{2\pi}} \exp\left[-1/2\left[\frac{x_{ik}-\bar{x}_{ik}}{u_{ik}}\right]^2\right]$$

If the same probability function $P(x_{jk},\bar{x}_{jk},u_{jk})$ can be written for the jth data vector, the overlap-integral distance between points i and j is:

$$d_{ij} = \sum_{k=1}^{m}\left[1-\left[\frac{u_{ik}u_{jk}}{u_{ik}+u_{jk}}\right]^{1/2} \int_{-\infty}^{\infty} P_iP_j dx\right]$$

where "m" is the number of features in the data matrix.

Gaussian Feature-Space Distance: The Gaussian feature-space calculation is a non-metric form of the distance. It assumes a Gaussian probability function for one data vector as determined by its uncertainties and calculates the distances relative to itself, only. If x_{ik} and x_{jk} are the kth feature of data vectors i and j respectively, and u_{ik} is the uncertainty in x_{ik} then a dimensionless range z_{ik} for the average of x_{ik} and x_{jk} can be calculated as:

$$z_{ik} = \frac{[(x_{ik} + x_{jk})/2] - x_{ik}}{u_{ik}}$$

The distance calculation, utilizing z_{ik} in determination of the area under the probability curve is:

$$d_{ij} = \sum_{k=1}^{m}\left[1 - \frac{2}{\sqrt{2\pi}} \int_{0}^{z_{ik}} e^{-\frac{x^2}{2}} dx\right]$$

This distance can be made metric by using the square root of the sum of the squares of the related uncertainties for a pair of features from the two data vectors for the average, instead of the two corresponding uncertainties.

Principal Component Factor Analysis

Within Arthur, there exists a subsystem of mutually independent routines designed to aid in the reduction of the dimensionality of the data by eigenanalysis (principal component factor analysis). The goal of these methods is the extraction of as much variance as possible in a minimum number of features by a method of analysis which will yield relatively stable and invariant results. This is achieved by extraction of the minimum number of factors (eigenvectors) which best span the original data. Several routines are provided which aid in the analysis of the factors derived. The following is a brief description of each subroutine of the subsystem.

KAPRIN. The extraction of the eigenvectors and eigenvalues of a data dispersion matrix is performed in KAPRIN (19). The routine begins by forming the n x n dispersion matrix "D" from the input data such that

$$D = X^T X$$

where X is an m x n matrix of data vectors. The characteristic eigenvectors and eigenvalues of the dispersion matrix are found by the Jacobi diagonalization method and ordered from highest eigenvalue to lowest.

KATRAN. Creates a new data matrix from the first k factors of the data (19). If X' is the data matrix formed by the factor matrix, V, from a data matrix X, then

$$X' = XV^T$$

In cases where the uncertainty values for the features are included with the data the related uncertainties are transformed into a matrix along with the sample matrix.

KACROSS. A convenient measure of how well a given feature aligns with the variance spanned by a given factor is provided by the squared correlation of that feature with each feature of the data transformed by the factor (9). If X' is a matrix to which X is being compared, then the square correlation of the jth feature of X to the kth feature of X' is

$$C^2_{j,k} = \frac{\left[\sum_{i=1}^{m} (x_{i,j}-\bar{x}_j)(x'_{i,k}-\bar{x}'_k)\right]^2}{\sum_{i=1}^{m} (x_{i,j}-\bar{x}_j)^2 \sum_{i=1}^{m} (x'_{i,k}-\bar{x}'_k)^2}$$

where "m" is the number of data vectors.

KAVARI, the varimax rotation used within ARTHUR, is based on the method of Kaiser (20). It consists of a large number of orthogonal transformations involving two factors at a time which maximize the variance of the squared elements.

If V is an n x n factor matrix (matrix of ordered eigenvectors) of X, then the communality can be defined as:

$$h_i = \sum_{j=1}^{n'} v_{ij} \sqrt{\lambda_j} = \sum_{j=1}^{n'} \alpha^2_{ij}$$

where v_{ij} is an element of V with associated eigenvalue λ_j. The varimax rotation is the orthogonal transformation of V which maximizes the function

$$\sum_{i=1}^{n'}\left[n\sum_{j=1}^{n}\left[\frac{\alpha_{ij}^2}{h_i^2}\right]^2-\left[\sum_{i=1}^{n}\frac{\alpha_{ij}}{h_i}\right]^2\right]$$

A factor matrix of X is input for use as the initial loadings. Beginning with the first two vectors, then the first and third, etc., each pair of vectors is orthogonally rotated until the criteria of variances of their squared elements cease to improve.

<u>KAMALIN</u>. The average error criteria can be used to determine the number of factors required to reproduce the data within the average root mean square data uncertainty. If $X' = XV^T$, then $Y=X'V'$, where V' is V with the lower n-p rows zeroed out, spans as much of the original variance as contained in the first n-p factors in V. The average error associated with the n-p factors is

$$\bar{e} = \left[\sum_{i=1}^{m}\sum_{j=1}^{n}(x_{ij}-y_{ij})^2\right]^{1/2}$$

where "n" is the number of features in X and "m" is the number of data vectors.

<u>KAPICK</u> selects variables based on their contribution to the variance of the vector with the largest eigenvalue. The feature whose contribution is largest is selected, then deleted from the data. The n-1 remaining features are orthogonalized from the chosen feature forming a new data matrix.

<u>KAORTH</u> orthogonalizes a rotated vector matrix V, such that

$$V' = V(V^TV)^{-1}$$

It is used to correct for the small buildup of nonorthogonality encountered in the iterative varimax rotation.

<u>KADISP</u> creates a data file whose elements are distances as defined by the data dispersion matrix (<u>9</u>). The square dispersion matrix $D = X^TX$ where X is the input data matrix is written to disk file as a square data matrix for subsequent analysis.

<u>KAVECTOR</u>. The vectors produced in KAPRIN or KAVARI can be analyzed in KAVECTOR. The analysis includes the total number of positive eigenvectors, their contribution to the total variance, the percent contribution to the vector by each feature, the number of eigenvectors with eigenvalues larger than the average eigenvalue and Bart-

lett's χ^2 (9) values for determination of the number of meaningful factors.

KARLOV. (Karhunen-Loève transform) KAPRIN followed by KATRAN on mean scaled data is often used as a method of reducing the dimensionality of a data set. The features derived from the K-L transform (22) are the best approximation in a least-mean-square error sense to the original data. Computationally, the transform consists of extraction of the characteristic eigenvalues and corresponding eigenvectors of the data covariance matrix. The vectors are ordered from highest to lowest eigenvalue. Since the variance spanned by each vector is its eigenvalue "λ_i" the variance preserved by each over the original can be expressed as a percent of the total variance,

$$\% = \left[\frac{\lambda_i}{\sum_j \lambda_j} \right] \times 100$$

After normalization of each vector to unity, the original data matrix X can be transformed by the vector matrix V to form a new data matrix X', where $X' = XV^T$. This can be thought of as an orthogonal rotation $X \rightarrow X'$ such that the first new feature contains the greatest variance in the data and each successive feature represents the maximum residual variance.

Because the information (in a variance sense) is preserved in as few features as possible the Karhunen-Loève transform is often utilized for projection of the data into two or three dimensions for display.

Statistical Isolinear Multicategory Analysis (SIMCA)

SIMCA (21) is a subsystem made up of five independent routines to facilitate disjoint principal component analysis of categorized data. One model per category is built by principal component analysis of its data vectors. The optimal number of components for invariant description of each category can be determined or preselected. Classification of an unknown into a particular category is based on the similarity of the vector to the principal component model which represents the class.

Two methods are available for component extraction; the iterative NIPALS and the Jacobi diagonalization method. Optimization of the number of components is done by cross-validation.

SICSVA. Cross validation is used in SICSVA (21) to determine the optimal number of stable components necessary to describe the data. Each category is divided into T groups where T=(number of data vectors)/10. The first group of data vectors is deleted and a similarity model (see SIPRINCO) calculated for the remaining data. The deleted data vectors are fitted to the model and the residual deviations (see SICLASS) determined for each of "A" component models;

($A = 0,1,\ldots,n-2$, where n is the number of data vectors) or ($A = 0,1,\ldots,M-2$, where M is the number of variables). The second group is then deleted and compared to the model of the resulting data. This is done for all groups. For each value of A, the sum deviation, D_A over all T groups is determined. By making F-tests on $(D_{A-1}-D_A)/n$ versus $D_A/[n(M-A-1)]$ a determination of whether the last component added was significant or not can be made.

SIUTIL is a utility routine for output of the computed components for each category.

SIJACOBI. After extraction of feature means as a first component, SIJACOBI uses the Jacobi method for extraction of the eigenvectors of each category. Whereas teh NIPALS method can be used to calculate any number of components, the Jacobi method calculates all eigenvectors of the data. Therefore, the optimal method depends on the number of features in the data set and the number of components under consideration. Results of SIJACOBI and SIPRINCO are identical. For one or two components, SIPRINCO is faster. For more than two components, SIJACOBI is preferred.

SIPRINCO computes the similarity mode for each class by the NIPALS method (21). After extraction of feature means, each eigenvector and eigenvalue is computed one at a time by minimization of the residual variances within a category. If $Y_{i,k,m}$ is the ith feature of the kth data vector of category m, then $Y_{i,k,m}$ can be linearly related to a number (A) of sets of values for the parameters $\beta_{i,a,m}$ and $\Theta_{a,k,m}$ such that

$$Y'_{i,k,m} = Y_{i,k,m} - \alpha_{i,m} = \sum_{a=1}^{A} \beta_{i,a,m}\Theta_{a,k,m} + \varepsilon_{i,k,m}$$

where $\alpha_{i,m}$ is the mean of feature i in category m. The method begins by loading one of the data vectors of a category as an estimate of the Θ's. The "i" β's are then computed such that the residual variance for Y' is a minimum:

$$\sum \varepsilon^2_{i,k} = \min \rightarrow \beta_{i,m} = \frac{\sum_k \Theta_{k,m} Y'_{i,k,m}}{\sum_k \Theta^2_{k,m}}$$

the resulting β's are then used as estimates to minimize the Θ's:

$$\sum \varepsilon^2_{ik} = \min \rightarrow \Theta_{km} = \frac{\sum_i \beta_{im} Y'_{i,k,m}}{\sum_i \beta^2_{im}}$$

The process of estimation followed by minimization continues until it converges for some set of β_k and Θ_i. New Y' values are computed as $Y'_{ik} = Y'_{ik} - \beta_{im}\Theta_{mk}$ and the process begins again. This process continues until all A components have been determined.

SICLASS (21) classifies unknown data vectors on the basis of how well the principal component model for a class fits the data. If Y_i is the kth unclassified data vector, it is classified in the following way: The data vector Y_i is fitted to the parameters α_{im} (category means) and $\beta_{i,a,m}$ (principal components) for category m. The fitting corresponds to a linear regression

$$Y'_i = Y_i - \alpha_{im} = \sum_{a=1}^{A} \beta_{iam}\Theta_{ak} + \varepsilon_{im}$$

The residual variance,

$$s_m^2 = \sum_{i=1}^{n} \varepsilon_{im}^2/(n-A),$$

where n is the number of features, is a convenient measure of how well the data vector fits the class model m. If s_o is the standard deviation of the q objects in the reference class where

$$s_o^2 = \sum_{k}^{q} \sum_{i}^{n} (\varepsilon_{ik})^2/[(q-A-1)(n-A)]$$

then $s_m >> s_o$ (F-test) implies the data vector is not a member of the mth category.

Unsupervised Clustering

HIER is an unsupervised learning (cluster analysis) method based on the relative similarity of a set of data vectors (22). Each vector is initially assumed to be a lone cluster. A similarity matrix is constructed such that if $S_{i,j}$ is the similarity between the ith and j^{th} data vector, then

$$S_{i,j} = 1 - \left[\frac{d_{i,j}}{d_{max}}\right]$$

where $\frac{d_{i,j}}{d_{max}}$ is the interpattern distance of data vectors "i" and "j" normalized by the largest interpattern distance d_{max} in the data (see DISTANCE).

The matrix is scanned for the maximum similarity in the set. These "most similar" vectors are clustered, removed from the matrix and replaced by a new vector whose location is the average of the two vectors. In combining clusters, two options are available.

Either the average of the two clusters is weighted by the number of data vectors in each cluster or each cluster is given equal weight. The new matrix is scanned for the next greatest similarity and the procedure is repeated. The process ends when all the data vectors form a single cluster. Output is in the form of a connection dendrogram.

TREE is an unsupervised learning (cluster analysis) method that generates a minimal spanning tree (23) over the data vectors in the training set. The tree is then pruned to determine the "natural" clustering of the data. The tree is formed by connecting the data vectors (represented by data points in the feature space) such that each point forms a node of the tree (is attached by at least one line segment or edge) and the sum of these line segments over the entire data is a minimum. Clusters are pruned from the tree on the basis of the length of a line segment relative to the nearby line segments. If this edge is greater than some factor times the length of the nearby edges or more than a chosen deviation larger than the average length of the nearby edges, it is ruled inconsistent and the tree is clipped at this node. This routine, like HIER, looks for natural groupings or clusters in the feature space.

Non-Linear Mapping (24)

NLM is a display technique that preserves interpoint distances (see DISTANCE). The program is initiated with either a random vector or the mapping resulting from linear projection of the first two eigenvectors of the data (see K-L). The plotting distance is defined as the interpoint Euclidean distance in the two dimensional projection plane. If the true distance in n space is the Euclidean distance taken over all n variables then minimizing the error in the distance encountered in the two-dimensional mapping will result in the projection that best preserves the interpoint distance. A weighting factor is sometimes applied to preserve either large distances or small distances at the expense of each other. Since this minimization involves iteratively changing the positions of the points in two-space, a conjugate gradient technique is employed for the minimization.

Classification Based on Interpattern Distance

KNN classifies unknown data vectors on the basis of interpattern distances to data vectors in the training set (25). The distances between a given data vector and all other vectors of the training set data are ordered from smallest to largest. An "unknown" is classified by committee vote into that category which has the majority of k nearest neighbors. In case of a tie the vector is assigned to the category whose sum distance over k neighbors is the smallest.

PNN calculates the properties for continuous property data vectors based on the p-nearest neighbors of each vector. The criterion used for nearest is the interpattern distance (see DISTANCE). Seven property estimates are calculated such that if p_j is the property of the jth data vector of the m nearest neighbors to vector i, then

$$P_i = \sum_{j=1}^{m} P_j \,/m$$

where m is set to seven different preselected percents of the total data vectors in the data set.

Feature Selection by Weighting

WEIGHT (26,27) is a preprocessing method that weights each feature on the basis of its individual importance to the solution of a pattern recognition problem. For categorized data, the criterion of importance can be either the total variance or total Fisher weight for the feature. The variance weight is a ratio of the interclass variance of two categories to the intraclass variances of the categories. If $W_{j,m,n}$ is a measure of the utility of feature j in separating categories m and n, the variance weight $(WV)_{j,m,n}$ is:

$$(WV)_{j,m,n} = \frac{\sum_{k=1}^{Nm} \frac{x_{k,m,j}^2}{Nm} + \sum_{k=1}^{Nn} \frac{x_{k,n,j}^2}{Nn} - \frac{2\sum_{k=1}^{Nm} x_{k,m,j} \sum_{k=1}^{Nn} x_{k,n,j}}{NmNn}}{2\left[\sum_{k=1}^{Nm} \frac{(x_{k,m,j}-\bar{x}_{m,j})^2}{Nm} + \sum_{k=1}^{Nn} \frac{(x_{k,n,j}-\bar{x}_{n,j})^2}{Nn}\right]}$$

where N_i is the number of data vectors in category i; the total variance weight is the geometric mean of the individual category pair weights. The Fisher weight is a ratio between the square difference in the category pair means and the sum of intraclass variances:

$$(WF)_{j,m,n} = \frac{(\bar{x}_{m,j} - \bar{x}_{n,j})^2}{\sum_{k=1}^{Nm} \frac{(x_{k,m,j}-\bar{x}_{m,j})^2}{Nm} + \sum_{k=1}^{Nn} \frac{(x_{k,n,j}-\bar{x}_{n,j})^2}{Nn}}$$

The total Fisher weight is the arithmetic average of the individual category pair weights.

For continuous property data the weighting in done on the basis of the correlation of the feature to the property. The square correlation to property of feature j is:

$$\frac{\left[\sum_{k=1}^{N}(x_{j,k}-\bar{x}_j)(p_k-\bar{p})\right]^2}{\sum_{k=1}^{N}(x_{j,k}-\bar{x}_j)^2\sum_{k=1}^{N}(p_k-\bar{p})^2}$$

where N is the number of data vectors in the training set and p_k is the property of the kth data vector.

SELECT (28) is a feature selection technique that generates orthogonal features based on their importance to classification. The criterion for importance for categorized data is the variance or Fisher weight and for continuous-property data, the correlation-to-property weight (see WEIGHT). The highest weighted feature is selected as the first feature. The remaining features are then decorrelated from the chosen feature. The decorrelated features are reweighted and the feature whose new weight is highest becomes the second selected feature. The process continues until either a specified number of features is chosen or a given minimum weight attained. The selected (unweighted) features are output to a file for later use. The user can opt for the decorrelated features or the same features in their unchanged form. Since one set is a linear combination of the other set, the same information is retained for either option. Only the representation is changed (i.e. the sub-feature space is either rotated or not rotated to orthogonal axes).

GRAB. As a feature selection method, GRAB (12) is intermediate between weight (with no feature decorrelation) and the more expensive SELECT (with total decorrelation). A previously-weighted file of n data vectors is input to the routine. Each feature is assigned an initial weight

$$W(1)_i = \left(\sum_{k=1}^{n}(x_{i,k}-x_i)^2\right)^{1/2}$$

The feature with the largest weight is selected as the first new feature. Each of the remaining features is reweighted such that if $C_{i,j}$ is the correlation between the ith feature just chosen and the remaining feature j,

$$W(2)_j = W(1)_j[1-|C_{i,j}|]$$

For the mth iteration the weight of the jth feature remaining is

$$W(m)_j = W(1)_j\prod_{i=1}^{m-1}[1-|C_{i,j}|]$$

the resulting selected features are autoscaled but neither weighted nor decorrelated.

SCALE offers several methods of scaling features (26) in the data. The scaling factors are derived from the n data vectors of the training set and applied to all of the data. For data for which the uncertainties are known, several scaling schemes are available. The following conventions are available for scaling without uncertainty weighting:

Autoscaling: If $X_{i,j}$ is the ith feature associated with the jth data vector, then

$$x_{i,j} = \frac{(x_{i,j}-\bar{x}_i)}{\left[\sum_{j=1}^{n} (x_{i,j}-\bar{x}_i)^2\right]^{1/2}}$$

where n is the total number of data vectors in the training data. The resulting new features all have a mean of 0.0 and a variance of 1.0. This removes any inadvertent weighting that might occur due to the difference in magnitude of the features.

Range scaling: If $Xmin_i$ and $Xmax_i$ and the minimum and maximum values respectively of feature i in the training data then

$$x_{i'k} = \frac{(x_{i,k}-Xmin_i)}{(Xmax_i-Xmin_i)}$$

scales each feature to a range of 1 lying between 0.0 and 1.0.

Mean subtraction:

$$x_{i'k} = (x_{i,k}-\bar{x}_i)$$

Variance normalization:

$$x'_{i,k} = \frac{x_{i,k}}{\sum_{j=1}^{n} x^2_{i,j}}$$

Mean normalization:

$$x'_{i,k} = \frac{x_{i,k}}{\bar{x}_i}$$

Scaling methods which weight the measurement by its uncertainty (3) include: (a) Error-weighted autoscale, (b) Error-weighted mean subtraction and (c) Error-weighted mean normalization. In each of these methods the mean in equation 1, 3, and 5 is replaced by a weighted mean, $\bar{x}_i$, which is calculated as:

$$\bar{x}_i = \frac{\sum_{j=1}^{n} \left[\frac{x_{i,j}}{u_{i,j}^2}\right]}{\sum_{j=1}^{n} \left[\frac{1}{u_{i,j}^2}\right]}$$

where the sum is over the data vectors and $u_{i,j}$ is the uncertainty associated with feature $x_{i,j}$.

Linear Discriminant Analysis

The linear classification section of ARTHUR contains routines for both category and continuous property data. For category data, multi-linear regression and hyperplane discriminant analysis are available.

LEAST performs a lest-squares multi-linear regression (29) that is best suited to continuous property problems. If D is a data matrix with associated property matrix P, then $W=(D^TD)^{-1}D^TP$ is the least squares solution to the set of linear equations P=DW where W is a vector which weights the utility of the features in fitting the data.

In actual practice, determination of the weight vector is done by

$$W = [E^TC^{-1}E]X^TP$$

where X is obtained by mean normalization of D, C^{-1} is the inverted correlation matrix associated with D and E is a diagonal matrix whose elements are the reciprocal variances of the features.

Prediction of an unknown property P' is based on the weight vector obtained is therefore

$$P' = X'W$$

LEDISC is a multi-linear least squares regression designed for categorized data. Except in property definitions it is computationally equivalent to LEAST. For a data set of n categories, n linear regressions are performed such that for the ith regression the property P is defined as

$$P = \begin{cases} +1 \text{ for all vectors in category } i \\ 0 \text{ for all vectors not in category } i \end{cases}$$

An unknown data vector is placed into that class whose weight vector produces the largest value.

LESLT is a variable reduction technique which seeks to optimize category pair separation in as few variables as possible (30). A feature derived is a linear combination of the original data that describe the position of a data vector relative to a hyperplane between two categories in the data set. The input data matrix (X) of n categories is divided into n(n-1)/2 submatrices. If Y is the submatrix containing only those patterns in categories i and j plus the test data, an outcome column matrix of properties can be defined such that

$$G^{i,j} = \begin{cases} -1 \text{ for patterns in } i \\ +1 \text{ for patterns in } j \end{cases}$$

Thus defined, there exists a vector W_k of weights such that $YW_k = G^{i,j}$. (Determination of W_k is the least squares solution for this equation (see LEAST).) The weight vector obtained is used to transform and classify all the data vectors in Y. This process is followed for all category pairs. Once all the weight vectors are obtained, the entire data matrix (X) is transformed such that $X' = XW$. The new matrix X' obtained has n(n-1)/2 features which are approximate category-pair separators.

LEPIECE (12) does a piece-wise least squares multiple regression for each data vector in the training and test set. The property of each data vector is predicted from the fit (see LEAST) using the k-nearest-neighbors (see KNN) to the vectors. The value of k is a user-defined multiple of the number of features. The criterion used for "nearest" is the interpattern distance (see DISTANCE). Only those features used in the determination of the distance are used in the regression.

MULTI is a hyperplane discriminant function method designed for multi-category data (31). Computationally, it is equivalent to PLANE, except in category definition. For a data matrix of n categories, n hyperplanes are generated such that the ith hyperplane describes the separation of the ith category from the rest of the data.

PLANE generates and classifies on the basis of a linear discriminant function (31) and is best suited to data containing two categories (see MULTI for multicategory case). By an error-correction feedback method it seeks a hyperplane in an augmented n+1 space (where n is the number of features) that best separates a pair of categories.

Each data vector in n space is considered a vector in n+1 space where the $n+1^{th}$ feature is unity. Therefore, two classes can be defined as lying on either side of a hyperplane (whose equation in n+1 space is $W \cdot Y = \emptyset$), through the origin with corresponding class numbers +1 and -1. The discriminant function is calculated by first loading a weight vector W with random or user-defined values. During training, classification of vector Y_k by this weight vector is a decision of the form

$W \cdot Y_k = S_k =$ correct, if the sign of the response relative to the hyperplane is the same as the sign of its class

incorrect, if the sign is not the same

If a pattern is misclassified, the weight vector is adjusted by reflection of the hyperplane about the misclassified point. The new weight vector is then used to classify the data. The process continues until all patterns in the training set are correctly classified or a maximum number of iterations is reached.

For more than two categories, a hyperplane separating each pair of categories is found. An unknown data vector is then classified using a majority committee vote procedure on all the discriminant function responses. The use of PLANE for multi-category data is equivalent to a piece-wise learning machine (31).

REGRESS is a multidimensional multivariate regression method which computes a linear discriminant function. It accepts both category and continuous data. Two optimization methods are available. Either the residual variance or the multiple correlation can be minimized.

STEP (32) is a stepwise multi-linear regression method. Features used in the regression are determined by their contribution to the overall variance. In the regression, features are added one at a time such that the feature that is added makes the greatest improvement in the "goodness of fit." When a feature that is indicated to be significant to the reduction in variance in an early stage of the regression is indicated to be insignificant after the addition of several other features, it is eliminated from the regression before addition of another feature. The criterion for selection of a feature to add or remove from the calculation is as follows:

Removal: If the variance contribution is insignificant at a specified F-level, the feature is removed from the regression.

Addition: If the variance reduction due to addition of a feature is significant at a specified F-level, this feature is entered into the regression.

<u>Utilities</u>

The following is a series of routines that permit the user to easily control processing of the data. A major source of ARTHUR's versatility is provided through the "CHANGE" (CHXXX) and the "TUNE" (TUXXX) routines. These routines allow the user to change the various definitions of a problem without changing the form of the input data. The Calcomp and Tektronics plot routines are designed to run on the CDC. The printer/plotter routine VARVAR is machine independent. Any data file can be listed or punched at any point in a run by a call to UTILIT.

<u>NEW</u> initializes the program. System files are rewound and all except one are initialized by writing a one record header at the beginning of the file. Several fixed common parameters can be redefined with a call to this routine.

<u>ENDIT</u> terminates the program. Termination by this routine may occur by a user call or by the encounter of a recognizable error condition during a run.

<u>UTILIT</u> provides a line printer listing of the data matrix and/or the distance matrix.

<u>INPUT</u>. In INPUT, a coded data matrix may be input to the program. If the data are categorized, it is reordered such that all data vectors belonging to the same class occur together. For continuous data, the user may opt for reordering by the magnitude of the property. Missing values in the data are flagged with a value equal to the largest real number allowed in the program. The data matrix is output to a binary file that is compatible with all other routines in ARTHUR.

<u>INFILL</u> fills in any missing data in the data matrix. For category data, a missing feature value in a data vector of the training set is filled with the mean value of the feature for the category to which the vector is a member. A missing feature value in the test set is filled with the mean value of the feature for all the training data. For continuous property data the feature value is filled with the mean of the data.

<u>INDUMP</u> deletes constant and redundant features in the data. For category data, the occurrence of a constant feature in a category results in an automatic call to terminate the program since many methods employ variance in the data features as a criterion.

<u>VARVAR</u> produces line printer plots of a data matrix. Two options are available. Either two features may be plotted against each other or one feature may be plotted against the properties of the data vectors.

VACALC produces calcomp plots equivalent to the line printer plots available in VARVAR.

VATKIT produces plots for a Tektronics graphics terminal. These plots are equivalent to those produced in VARVAR.

CHANGE allows the user to quickly and conveniently change a data matrix definition from continuous property to category and vice versa.

CHCATEGORY. In CHCATEGORY the user can redefine the category of selected data vectors, reorder the data, or create a new data matrix with a selected number of categories retained in the training set and all others placed in the test set.

CHDATA allows manipulation of the data vectors in a data matrix. Specified vectors may be moved between the training and test sets, and vectors may be deleted from the data.

CHFEATURE provides feature manipulation. Features may be deleted from the data, transformed (by addition, subtraction, multiplication, division, exponentiation, and logarithmic substitution) and/or combined by any of these operations to form new features.

CHJOIN combines the matrices of two data files. This can occur in two modes. Either the data vectors of the files are connected or the features of the two files are combined. In either case, a new data file is created from the merging.

CHSPLIT. User defined categories may be split off onto an alternative data file in CHSPLIT.

CHSUB creates a new data file by randomly selecting a subset of the data vectors in the data matrix. By default all categories in the data retain 80% of their data vectors. This percent can be redefined by the user.

CHUNCE. Feature uncertainties can be added, changed, or deleted from the data file in CHUNCE. The uncertainty may be added as a relative or absolute error. For a given feature (i) of a given vector (j) the uncertainty $u_{i,j}$ is defined as

$$u_{i,j} = (xabs)_{i,j} + \frac{(xrel)_{i,j} * x_{i,j}}{100}$$

where $(xabs)_{i,j}$ is the absolute error and $(xrel)_{i,j}$ is the relative error in percent of the feature $x_{i,j}$. By default, $(xabs)_{i,j} = 0$, and $(xrel)_{i,j} = 10.0\%$.

TUNE generates a new data file with features formed from the n(n-1) ratios of the original measurements. This simple transform not only offers more features to the various selection algorithms, but also, in many cases, lend stability to the features. For example, in cases where inert methods for sample dilution have taken place before measurement, the ratios of concentrations of elements in a group of samples more readily lend themselves to establishing a common origin than the individual concentrations (18).

TURAND perturbs each measurement of each vector by a function of the error associated with the feature. If $x_{i,j}$ is the ith feature of the jth data vector, then the error perturbed feature $x'_{i,j}$ is:

$$x'_{i,j} = x_{i,j} + \alpha_{i,j}s_{i,j}$$

where $s_{i,j}$ is the standard deviation of the distribution of $x_{i,j}$ and $\alpha_{i,j}$ is a random number from a gaussian distribution of unit variance and zero mean (33).

TUMED normalizes all vectors in a data matrix by their Euclidean distances from a defined origin. This transform is often useful in data that exhibit a "time" dependence.

TUTRAN takes the transpose of a data matrix. The new data file that is formed is output to a disk file for subsequent analysis in other routines.

Acknowledgment

We would like to thank Maynarhs Da Koven for his help in creating ARTHUR. Special thanks are given to Robert W. Gerlach for help and criticisms on this manuscript.

Literature Cited

1. Duewer, D. L., Harper, A. M., Koskinen, J. R., Fasching, J. L. and Kowalski, B. R., ARTHUR, Version 3-7-77.
2. Koskinen, J. R. and Kowalski, B.R., Journal of Chemical Information and Computer Science (1975), 15, 119.
3. Fasching, J. L., Duewer, D. L. and Kowalski, B. R., submitted to Analytical Chemistry.
4. Nie, N. H. et al, "Statistical Package for the Social Sciences, 2nd Edition" McGraw Hill, Inc., New York, 1975.
5. Dixon, W. J., Ed., "BMD, Biomedical Computer Programs," University of California Press, Berkeley, 1971.
6. Kowalski, B. R. and Reilley, C.A., Analytical Chemistry, (1971), 42, 1387.
7. Kowalski, B. R. and Bender, C. F., Analytical Chemistry, (1973), 45, 2334.
8. Miller, R. G., Biometrika, (1974) 61, 1.
9. Duewer, D. L., et al., Analytical Chemistry (1976) 48, 2002.

10. Kowalski, B. R. et al., Analytical Chemistry (1972) 44, 2176.
11. Stevenson, D. F. et al., Archaeometry (1971) 13, 17.
12. Duewer, D. L. et al., "Documentation for ARTHUR, Version 1-8-75," (1975) Chemometrics Society Report No. 2.
13. Reinsch, C. H., Numerische Mathematik (1967) 10, 177.
14. Birnbaum, Z. W., JASA (1952) 47, 425.
15. Davies, O. L. and Goldsmith, P. L., "Statistical Methods for Research and Production," p. 234, Hafner, New York, 1972.
16. Anderson, T. W., "An Introduction to Multivariate Statistical Analysis," p. 65, John Wiley & Sons, New York, 1958.
17. Mahalanobis, P. C., Proceedings of the National Institute of Science of India, p. 49, 122, 1936.
18. Anders, O. U., Analytical Chemistry (1972) 44, 1930.
19. Any numerical analysis text.
20. Horst, P., "Factor Analysis of Data Matrices," Holt, Rinehart and Winston, Inc., New York, 1965.
21. Wold, S., Journal of Pattern Recognition (1976) 8, 127.
22. Kowalski, B. R. in "Computers in Chemical and Biochemical Research, Vol. 2," Academic Press, New York, 1974.
23. Andrews, H. C., "Introduction to Mathematical Techniques in Pattern Recognition," Wiley Interscience, New York, 1972.
24. Kowalski, B. R. and Bender, C. F., Journal of the American Chemical Society (1973) 95, 686.
25. Cover, T. M. and Hart, P. E., IEEE Transactions on Information Theory, IT-13, 21 (1967).
26. Kowalski, B. R. and Bender, C. F., Journal of the American Chemical Society (1972) 94, 5632.
27. Fisher, R. A., Annals of Eugenics (1936) 7, 179.
28. Kowalski, B. R. and Bender, C. F., Journal of Pattern Recognition (1976) 8, 1.
29. Kowalski, B. R., Snalytical Chemistry, (1969) 41, 695.
30. Kowalski, B. R. and Bender, C. F., Analytical Chemistry (1973) 45, 590.
31. Nilsson, N. J., "Learning Machines," McGraw-Hill, New York 1965.
32. Ralston, A., and Wolf, H. S., "Mathematical Methods for Digital Computers," p. 191, John Wiley and Sons, New York, 1966.
33. Hamming, R. W., "Introduction to Applied Numerical Analysis," McGraw-Hill, New York, 1971.

3

Abstract Factor Analysis—A Theory of Error and Its Application to Analytical Chemistry

EDMUND R. MALINOWSKI

Department of Chemistry and Chemical Engineering, Stevens Institute of Technology, Hoboken, NJ 07030

Factor analysis (FA), a computer method for solving multi-dimensional problems, is rapidly gaining importance in many phases of chemistry (1-5). The first step in FA is concerned with determining the number of factors which are responsible for the data. Unfortunately, because of experimental uncertainty this is not an easy task. Most FA methods depend upon an accurate estimate of the uncertainties present in the data. Even then, each FA method may yield a different estimate of the size of the factor space. This poses a fundamental dilemma to the factor analyst in the early stages of his analysis.

In the present paper we will develop a theory of error which helps us overcome this difficulty. Our approach will be quite different from the traditional statistical methods reported by others (3,6). Our attention will be focused on how the error mixes into the abstract factor analysis (AFA) scheme. AFA is that part of the overall FA process which is concerned with data reproduction using the abstract mathematical factors produced by the decomposition of the covariance matrix. The theory shows that the eigenvalues can be grouped into two sets: a primary set which contains the true factors together with a mixture of error and a secondary set which is composed of pure error. By deleting the secondary set we actually remove error. Consequently the AFA reproduced data is a better representation of the real data than the raw experimental data used in the analysis. Although this is not the prime purpose of AFA, it does constitute an unexpected and useful fringe benefit.

The theory shows that three types of error (real error, RE; extracted error, XE; and imbedded error, IE) exist. These errors are mutually related in a pythagorean sense and can be calculated from the secondary eigenvalues, the size of the data matrix and the size of the factor space.

Arguments are presented to show how the IE function can be used to determine not only the dimension of the factor space but also the real error. Most importantly, this is accomplished without recourse to a knowledge of the experimental error.

An empirical function, called the factor indicator function, IND, is also described. This function, similar to the IE function, is calculated from the secondary eigenvalues. However, it is more sensitive than the IE function in determining the size of the factor space.

Mathematical models are used to illustrate and confirm the theory. The method is then applied to nuclear magnetic resonance, absorption spectrophotometry, mass spectra, gas-liquid chromatography and drug activity.

Theory of Error

Factor analysis is concerned with a matrix of data points. It is based upon expressing each raw data point d_{ik} as a linear sum of product terms. If the data contained no experimental error, the value of a pure data point d^*_{ik} would be expressed as follows:

$$d^*_{ik} = \sum_{j=1}^{j=n} r^*_{ij} c^*_{jk} \quad (1)$$

where r^*_{ij} is the jth cofactor of the ith row designee and c^*_{jk} is the jth cofactor of the kth column designee. The sum is taken over all n factors which are responsible for the data.

However, because of experimental error each raw data point is best represented as a sum of pure data and an error e_{ik},

$$d_{ik} = d^*_{ik} + e_{ik} \quad (2)$$

Such errors mix into the FA process, perturbing the cofactors and producing an excessive number of factors. Instead of eq. (1) we obtain eq. (3)

$$d_{ik} = \sum_{j=1}^{j=c} r^{\neq}_{jk} c^{\neq}_{jk} \quad (3)$$

where the sum is taken over c factors rather than n; c, of course, is greater than n. Superscript $\neq$ is used here to distinguish these cofactors from those belonging to the pure data.

From our knowledge of matrix algebra we can readily show that the number of terms in the above sum is either r, the number of rows in the raw data matrix, or c, the number of columns in the raw data matrix, whichever is the smaller number. Throughout our discussion we will assume that c is smaller than r. Hence the sum in eq. (3) involves c terms.

The proposed theory of error is based upon the following observation. Although c eigenvectors are required to span the total raw data space, only n of them are required to span the data space within experimental error. Because any orthogonal set of axes, of the proper number, can be used to describe the error space the same basis axes used to describe the raw data space can be used to describe the error space. Accordingly, then, the error e_{ik} associated with data point d_{ik} can be expressed as follows:

$$e_{ik} = \sum_{j=1}^{j=n} \sigma^{\neq}_{ij} c^{\neq}_{jk} + \sum_{j=n+1}^{j=c} \sigma^{o}_{ij} c^{\neq}_{jk} \tag{4}$$

Here $c^{\neq}_{jk}$ is a component of the jth basis axis used to describe the raw data, $\sigma^{\neq}_{ij}$ is projection of the ith error onto the jth axis and σ^{o}_{ij} is the projection of the ith error onto the jth axis of the pure error space. Notice here that we have grouped the error factors into two distinct sums. The first sum, from $j = 1$ to $j = n$, is associated with the important eigenvectors (called primary eigenvectors) which are required to account for the real data. The second sum, from $j=n+1$ to $j=c$, is associated with unnecessary eigenvectors (called secondary eigenvalues) which are produced by experimental error.

Equation (4) represents the foundation of the proposed theory of error. We will now explore its significance and confirm its validity by tracing its path through the factor analysis scheme.

Upon placing eq. (4) and (1) into (2) we find

$$d_{ik} = \sum_{j=1}^{j=n} (r^{*}_{ij} c^{*}_{jk} + \sigma^{\neq}_{ij} c^{\neq}_{jk}) + \sum_{j=n+1}^{j=c} \sigma^{o}_{ij} c^{\neq}_{jk} \tag{5}$$

Comparing this result with eq. (3) we see that

$$r^{\neq}_{ij} = r^{*}_{ij} \frac{c^{*}_{jk}}{c^{\neq}_{jk}} + \sigma^{\neq}_{ij} \tag{6}$$

In other words eqs. (3) and (5) can be expressed as

$$d_{ik} = \sum_{j=1}^{j=n} r^{\neq}_{ij} c^{\neq}_{jk} + \sum_{j=n+1}^{j=c} \sigma^{o}_{ij} c^{\neq}_{jk} \tag{7}$$

This can also be written as

$$d_{ik} = d^{\neq}_{ik} + e^{o}_{ik} \quad (8)$$

where

$$d^{\neq}_{ik} = \sum_{j=1}^{j=n} r^{\neq}_{ij}\, c^{\neq}_{jk} \quad (9)$$

and

$$\hat{e}^{o}_{ij} = \sum_{j=n+1}^{j=c} \sigma^{o}_{ij}\, c^{\neq}_{jk} \quad (10)$$

Since e^{o}_{ik} is associated with pure error the retention of an excessive number of eigenvectors simply tends to reproduce experimental error. Hence the secondary set of eigenvectors should be deleted. When this is done $d^{\neq}_{ik}$ will represent the difference between the raw data point and the reproduced data point.

The imbedded error $e^{\neq}_{ik}$ is defined as the difference between the reproduced data point and the pure data point:

$$e^{\neq}_{ik} = d^{\neq}_{ik} - d^{*}_{ik} \quad (11)$$

Because the trace of the covariance matrix, constructed by premultiplying the data matrix by its transpose, is invariant upon a similarity transformation, the following is true:

$$\sum_{i=1}^{i=r} \sum_{k=1}^{k=c} d^{2}_{ik} = \sum_{j=1}^{j=n} \lambda^{\neq}_{j} + \sum_{j=n+1}^{j=c} \lambda^{o}_{j} \quad (12)$$

where

$$\lambda^{\neq}_{j} = \sum_{i=1}^{i=r} \left(r^{*}_{ij} \frac{c^{*}_{jk}}{c^{\neq}_{jk}} + \sigma^{\neq}_{ij} \right)^{2} \quad \text{for } j = 1,\ldots,n \quad (13)$$

and

$$\lambda^{o}_{j} = \sum_{i=1}^{i=r} \sigma^{o}_{ij}{}^{2} \quad \text{for } j = n+1,\ldots,c \quad (14)$$

Here $\lambda^{\neq}_{j}$ is a primary eigenvalue and λ^{o}_{j} is a secondary eigenvalue. The largest eigenvalues contain the pure cofactors and thus belong to the primary set. The smallest eigenvalues contain nothing but pure error cofactors and thus belong to the secondary set. Deletion of the secondary eigenvalues and their associated eigenvectors from the AFA scheme should lead to data improvement. Equations (13) and (14) are important because they show how the error mixes into the eigenvalues.

We can also construct a covariance matrix from the AFA regenerated data matrix. Its trace is also invariant upon a similarity transformation. Hence

$$\sum_{i=1}^{i=r} \sum_{k=1}^{k=c} d^{\neq}{}_{ik}{}^{2} = \sum_{j=1}^{j=n} {}^{\neq}{}_{j} \qquad (15)$$

By subtracting eq. (15) from (12) we obtain

$$\sum_{i=1}^{i=r} \sum_{k=1}^{k=c} \left(d^{2}{}_{ik} - d^{\neq}{}_{ik}{}^{2} \right) = \sum_{j=n+1}^{j=c} \lambda^{o}{}_{j} = \sum_{i=1}^{i=r} \sum_{j=n+1}^{j=c} \sigma^{o}{}_{ij}{}^{2} \qquad (16)$$

In

In the discussion which follows we will use these equations to develop fully our understanding of the real error, extracted error, imbedded error, and their interrelationship.

Residual Standard Deviation - The Real Error

The residual standard deviation (RSD) is defined in terms of the projections of the error points onto the secondary axes, namely

$$r(c-n)(\mathrm{RSD})^{2} = \sum_{i=1}^{i=r} \sum_{j=n+1}^{j=c} \sigma^{o}{}_{ij}{}^{2} \qquad (17)$$

Because the secondary axes involve pure error components which are deleted from the reproduction we see that the RSD is, in reality, a measure of the real error (RE), the difference between the pure data and the raw data.

By placing eq. (16) into (17) we find

$$\mathrm{RE} = \mathrm{RSD} = \left[\frac{\sum_{j=n+1}^{j=c} \lambda^{o}{}_{j}}{r(c-n)} \right]^{1/2} \qquad (18)$$

Root-Mean-Square Error

The root-mean-square error is defined in terms of e^{o}_{ik}, the difference between a raw data point and its value regenerated by AFA, namely

$$rc(RMS)^2 = \sum_{i=1}^{i=r} \sum_{k=1}^{k=c} e^{o\,2}_{ik} \tag{19}$$

Because the reproduced data matrix is orthogonal to its associated residual error matrix,

$$\Sigma\Sigma \; e^{o\,2}_{ik} = \Sigma\Sigma \; d^2_{ik} - \Sigma\Sigma \; d^{\neq}_{ik} = \Sigma\Sigma \; (d_{ik} - d^{\neq}_{ik})^2 \tag{20}$$

Hence, we can readily see from (16), (19) and (20) that

$$RMS = \left[\frac{\sum_{j=n+1}^{j=c} \lambda^{o}_{j}}{rc} \right]^{1/2} \tag{21}$$

Pythagorean Relationship

Let us now theoretically form a covariance matrix from the experimental error matrix. The trace of this matrix is invariant upon a similarity transformation. Hence, recalling eq. (4), we conclude that

$$\sum_{i=1}^{i=r} \sum_{k=1}^{k=c} e^2_{ik} = \sum_{i=1}^{i=r} \sum_{j=1}^{j=c} \sigma^{\neq\,2}_{ij} + \sum_{i=1}^{i=r} \sum_{j=n+1}^{j=c} \sigma^{o\,2}_{ij} \tag{22}$$

Each of these three sums is intimately related to the residual standard deviation as follows:

$$rc(RSD)^2 = \sum_{i=1}^{i=r} \sum_{k=1}^{k=c} e^2_{ik} \tag{23}$$

$$rn(RSD)^2 = \sum_{i=1}^{i=r} \sum_{j=1}^{j=n} \sigma^{\neq\,2}_{ij} \tag{24}$$

$$r(c-n)\ (RSD)^2 = \sum_{i=1}^{i=r} \sum_{j=n+1}^{j=c} \sigma^{o}{}_{ij}{}^2 \tag{25}$$

Placing these three equations into eq. (23) we find

$$(RE)^2 = (IE)^2 + (XE)^2 \tag{26}$$

where

$$RE = RSD \tag{27}$$

$$IE = \left(\frac{n}{c}\right)^{1/2} (RSD) \tag{28}$$

$$XE = \left(\frac{c-n}{c}\right)^{1/2} (RSD) \tag{29}$$

The real error (RE) is the difference between the pure data and the raw data. The imbedded error (IE) is the difference between the pure data and the AFA reproduced data. The extracted error (XE) is the difference between the AFA reproduced data and the raw data. Figure 1 is presented as a mnemonic illustrating these relationships.

Several important facets of AFA now become apparent. Since $n < c$ we see clearly that IE < RE, hence AFA should always lead to data improvement. Secondly, by comparing eq. (29) to eqs. (18) and (21) we learn that the RMS, the difference between the AFA reproduced data and the raw data, is the extracted error. This is unexpected and quite surprising. If we use too many eigenvectors to reproduce the data we will reduce the extracted error and increase the imbedded error. It is important to use the proper number of eigenvectors in the reproduction process.

Testing the Theory with Mathematical Models

In order to test the proposed theory of error a series of mathematical models of various dimensionalities, sizes and errors were constructed and factor analyzed. The simplest and easiest model to visualize consisted of the one-dimensional data set depicted in Table I. The first two columns represent a pure data matrix. It is obviously one-dimensional since a plot of the points in column one against the corresponding points in column two produce a straight line of unit slope. When this pure data matrix is factor analyzed we obtain the results shown in the first column of Table II.

An artificial error matrix (third and fourth columns of Table I) was constructed and added to the pure data matrix to produce the raw data matrix shown in the fifth and sixth columns of

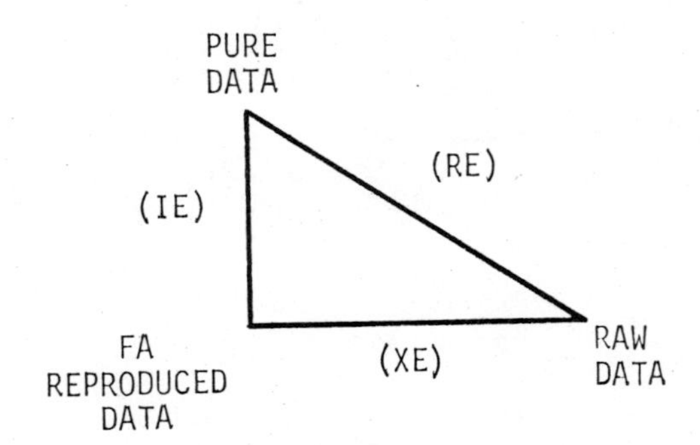

Figure 1. Mnemonic diagram of the pythagorean relationship between the real error RE, the extracted error XE, and the imbedded error IE, and their relationships to the pure, raw, and FA reproduced data

Table I - Artificial (One-Dimensional) Pure Data Matrix, Error Matrix, Raw Data Matrix and FA Reproduced Raw Data Matrix.

Pure Data Matrix $[D^*]$		Error Matrix $[E]$		Raw Data Matrix $[D] = [D^*] + [E]$		FA Reproduced Raw Data Matrix $[D^{\neq}]$	
column 1	column 2	column 1	column 2	column 1	column 2	column 1	column 2
1	1	0.2	0.0	1.2	1.0	1.0936	1.1052
2	2	-0.2	-0.2	1.8	1.8	1.7904	1.8095
3	3	-0.1	0.1	2.9	3.1	2.9846	3.0163
4	4	0.0	-0.1	4.0	3.9	3.9288	3.9705
5	5	-0.1	0.0	4.9	5.0	4.9240	4.9763
6	6	0.2	-0.2	6.2	5.8	5.9671	6.0305
7	7	0.2	-0.1	7.2	6.9	7.0118	7.0862
8	8	-0.2	0.1	7.8	8.1	7.9086	7.9926
9	9	-0.2	0.1	8.8	9.1	8.9033	8.9978
10	10	-0.1	0.2	9.9	10.2	9.9974	10.1036

Table I. This raw data matrix simulates real chemical data which contains experimental uncertainty. When this matrix was factor analyzed we obtained the results shown in the second and third columns of Table II. Two eigenvalues and two sets of row cofactors emerge from the analysis.

These cofactors are intimately related to the raw data points. To understand this relationship we plot the raw data points of column 1 against those in column 2. This plot is shown in Figure 2. Notice that the points lie in a two-dimensional plane. The row cofactors in Table II are the perpendicular projections of these points onto the primary and secondary axes which emerge from the FA. Notice that the projections onto the secondary axis contain nothing but error as predicted by the theory.

When the secondary eigenvector is deleted we obtain the FA reproduced raw data matrix shown in Table I. The RMS of the difference between the FA reproduced data and the pure data is calculated to be 0.087. The RMS of the errors listed in the error matrix is 0.148. This is in accord with our prediction based upon the proposed theory of error. The imbedded error, 0.087 is less than the real error, 0.148, by a factor of approximately $\sqrt{n/c} = \sqrt{1/2}$ as predicted by eq. (28). Exact agreement should not be expected because twenty data points do not constitute a good statistical sample.

Instead of a direct comparison we can calculate the RE and IE from the secondary eigenvalues via eqs. (18) and (28). These results are shown in Table III together with the direct calculations described above.

Many other artificial sets of data were constructed, having different sizes, dimensionalities and errors. A summary of the factor analyses of these matrices is given in Table III. In each case the RMS of the artificial error compares favorably with the real error predicted from eq. (18). Also the RMS of the difference between the pure and reproduced data compares favorably with the imbedded error predicted from eq. (28). These tests give credence to the proposed theory of error.

Imbedded Error Function

It is possible to deduce the true number of factors in a data matrix by studying the behavior of the IE as a function of n. If the errors are distributed randomly and fairly uniform throughout the data matrix, and are free from sporatic or systematic behavior, then we should expect their projections onto each of the secondary axes to be approximately the same. If this is true we may set $\lambda^o{}_j \cong \lambda^o{}_{j+1} \cong \ldots \cong \lambda^o{}_c$. Inserting this into eqs. (18) and (28) gives

$$IE = n^{1/2}k \text{ for } n > \text{true } n \qquad (30)$$

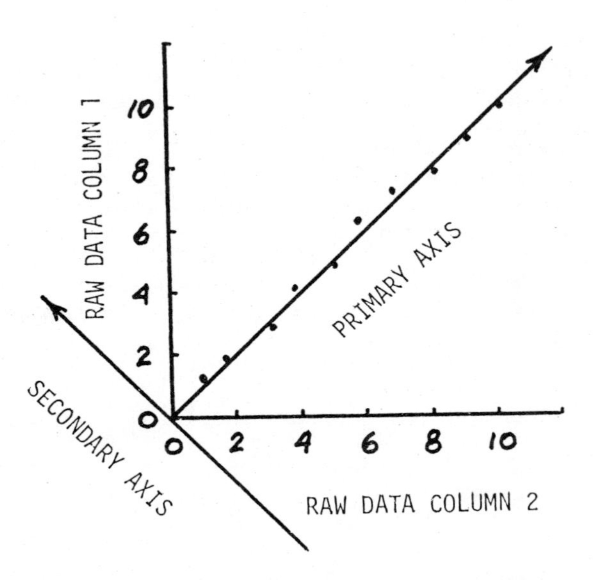

Figure 2. Illustration of the geometrical relationships between the raw data points of Table I and the primary and secondary axes resulting from factor analysis of the raw data matrix

Table II

EIGENVALUES AND ROW-COFACTORS RESULTING FROM FA

From Pure Data Matrix	From Raw Data Matrix	
λ_1	$\overline{\overline{\lambda}}_1$	λ^{o}_2
770	767.1514	0.2886124
r^*_{i1}	$\overline{\overline{r}}_{i1}$	σ^{o}_{i2}
1.41421	1.55487	0.14964
2.82843	2.54555	0.01344
4.24264	4.24333	-0.11901
5.65685	5.58569	0.10021
7.07107	7.00063	-0.03374
8.48528	8.48367	0.32765
9.89950	9.96895	0.26479
11.31371	11.24396	-0.15275
12.72792	12.65816	-0.14528
14.14214	14.21377	-0.13707

Table III - Summary of Model Data and the Results of Factor Analysis

Artificial Data and Artificial Error					Results of Factor Analyzing Impure Data Using n Factors		RMS of Difference between FA Reproduced Data and Pure Data $\left[\frac{\Sigma\Sigma(d_{ij}^{\neq} - d_{ij}^{*})^2}{rc}\right]^{1/2}$
n	r x c	Range of Pure Data	Range of Error	RMS	RE	IE	
1	10 x 2	1 to 10	-0.2 to 0.2	0.148	0.170	0.120	0.087
2	16 x 5	2 to 170	-0.08 to 0.08	0.041	0.041	0.026	0.026
2	16 x 5	2 to 170	-0.99 to 0.91	0.566	0.511	0.323	0.381
2	15 x 5	0 to 27	-1.9 to 2.0	1.040	0.962	0.608	0.726
3	10 x 6	0 to 32	-0.9 to 0.9	0.412	0.376	0.266	0.311
4	16 x 9	-581 to 955	-1.0 to 1.0	0.548	0.499	0.333	0.398
5	10 x 9	-137 to 180	-1.0 to 1.0	0.463	0.372	0.277	0.400

where k is a constant,

$$k = \left[\frac{\lambda^o_j}{rc} \right]^{1/2} \tag{31}$$

The imbedded error function should decrease as we employ more and more primary eigenvectors. However, when we use a secondary eigenvector in the reproduction scheme eq. (30) becomes effective. This equation shows that the IE will increase systematically as we use more secondary eigenvectors. The imbedded error function should reach a minimum when we use the exact number of eigenvectors. This minimum can be used to deduce the true factor space. Unfortunately, with real data this behavior will not always occur because the principal component feature of FA tends to exaggerate the nonuniformity of the error distribution and, in fact, places heavy emphasis on bad data points.

Factor Indicator Function

During our investigations we found an empirical function, which we call the factor indicator (IND), which is more sensitive than the IE function in determining the dimensionality of the factor space. This function involves the same variables (λ^o_j, r, c and n) and is defined as follows:

$$\text{IND} = \frac{\text{RE}}{(c-n)^2} \tag{32}$$

Similar to the IE function, the IND function reaches a minimum when the correct number of eigenvectors are employed.

Testing IE and IND Function Using Model Data

Prior to investigating real chemical data it is important to study the behavior of the IE and IND functions using artificial data, generated mathematically, for which we know every aspect of the pure data and errors introduced. Many such models were studied. Three typical examples are depicted in Table IV.

For model data A, the IE function decreases dramatically as we proceed from n = 1 to n = 4. The IE at n = 5 is larger than the value at n = 4 . This will occur when we have exceeded the true factor space. Hence we conclude that four factors are present. This conclusion is substantiated by the fact that the IND function shows a minimum at n = 4. Having concluded that there are four factors we then predict that the real error, RE, is approximately 0.50. These conclusions are in excellent agreement with the known facts. The pure data was generated by using four mathematical factors and the RMS of the random error added was 0.55.

For model data B the IE function shows a rapid drop from n = 1 to n = 5 and only a slight decrease from n = 5 to n = 6. This signifies that there are probably only five factors present. More dramatically, the IND function reaches a true minimum at n = 5, substantiating this conclusion. For n = 5 the RE is 0.37 in reasonable agreement with 0.46, the RMS of the error actually introduced.

It is important to recognize here that we have deduced the experimental error as well as the size of the factor space without relying upon any a priori knowledge of the error introduced.

Model data C was generated by haphazardly choosing numbers from 1 to 100. When this 10 x 8 data matrix was factor analyzed the results shown in Table IV were obtained. Notice that the IE increases as we proceed from n = 1 to n = 3. Notice also that the IND function increases from n = 1 to n = 7. Since the data is obviously not one dimensional it must be zero dimensional. In other words no factors are present. The data is not factor analyzable.

It is important to recognize here that we have discovered a criterion for deciding whether or not a data matrix is factor analyzable. This is so important that we recommend that it be used routinely as the first step in a factor analysis study.

Applications

The following sequence of studies was conducted using data which previously had been factor analyzed, the results having been reported in the scientific literature. The purpose was to see whether or not the newly-developed RE, IE and IND functions agreed with the previous conclusions and what new insight could be gleaned.

A. NMR Shifts. Weiner, Malinowski and Levinstone (2) conducted a FA study of the proton shifts of some 14 solutes in 9 common solvents. Using their data matrix we obtained the results shown in Table V. The IE gives evidence that only three factors are involved since only a very little decrease in this function, from 0.33 to 0.32, occurs on going from three to four factors. This conclusion is confirmed by the IND function which reaches a minimum at n = 3. For three factors the real error is predicted to be 0.58 Hz in excellent agreement with the experimental error which was estimated to 0.5 Hz. These conclusions agree with those of the original investigators.

When the 19F shifts reported by Abraham, Wileman and Bedford (7) was subjected to factor analysis we obtained the results shown in Table V. Both the IE and the IND functions give strong evidence that not two but three factors are operative.

Table IV - Results of Factor Analyzing Three Different Sets of Artificial Data

	Model Data A			Model Data B			Model Data C		
n	RE	IE	IND	RE	IE	IND	RE	IE	IND
1	140.21	46.74	2.1909	48.05	16.02	0.7507	27.00	9.54	0.55
2	46.27	21.81	0.9444	30.76	14.50	0.6277	24.66	12.33	0.69
3	1.95	1.13	0.0543	21.09	12.18	0.5859	21.88	13.40	0.88
4	0.50	0.33	0.0200	8.83	5.89	0.3532	18.51	13.09	1.16
5	0.45	0.34	0.0283	0.37	0.28	0.0233	14.42	11.40	1.60
6	0.37	0.30	0.0409	0.29	0.24	0.0323	11.02	9.54	2.75
7	0.32	0.28	0.0797	0.15	0.13	0.0376	6.47	6.05	6.47
8	0.25	0.24	0.2543	0.11	0.10	0.1084	-	-	-

A - A 16x9 four factor data matrix with values ranging from - 581.00 to 955.00 with an RMS error equal to 0.55.

B - A 10x9 five factor data matrix with values ranging from - 137.75 to 180.14 with an RMS error equal to 0.46.

C - A 10x8 data matrix consisting of random numbers ranging from 4 to 99.

Table V - Results of Factor Analyzing ^{1}H and ^{19}F Nuclear Magnetic Resonance Chemical Shifts of Various Solutes in a Variety of Solvents.

	PROTON SHIFTS[a]			FLUORINE SHIFTS[b]		
n	RE (Hz)	IE (Hz)	IND x 10^2	RE (ppm)	IE (ppm)	IND x 10^3
1	2.32	0.77	3.62	0.767	0.271	15.6
2	1.12	0.53	2.29	0.077	0.038	2.1
3	0.58	0.33	1.60	0.035	0.021	1.4
4	0.48	0.32	1.93	0.027	0.019	1.7
5	0.37	0.27	2.30	0.021	0.016	2.3
6	0.29	0.24	3.26	0.015	0.013	3.7
7	0.27	0.24	6.79	0.014	0.013	14.1
8	0.22	0.21	22.43			

a) Proton shifts (relative to internal TMS) of 14 solutes in 9 solvents. Data matrix was identical to Table I of Weiner, Malinowski and Levinstone, J. Phys. Chem., 74, 4537 (1970). The error was reported to be approximated 0.5 Hz.

b) Fluorine shifts (corrected for bulk-susceptibility effects using an external standard) of 14 nonpolar solutes in 8 nonpolar solvents. Data was taken in part from Table III of Abraham, Wileman and Bedford, J.C.S. Perkin II, 1027 (1973). The following solutes were deleted from the data matrix: $C_6H_5CF_3$, $CF_3CHC\ell Br$ and C_6F_{14}. The error was reported to be approximately 0.035 ppm.

Furthermore, for three factors, the RE value 0.035 is in perfect agreement with the reported error.

Without recourse to FA, Abraham and co-workers assumed that only two factors, the gas-phase shift and the van der Waals interaction, were responsible for the solvent shifts. FA gives clear evidence that a third factor, although small, makes a measurable contribution. Target transformation FA should be helpful in identifying the physical significance of this elusive factor.

B. Spectrophotometric Absorbances. Bulmer and Shurvell (8) factor analyzed the infared spectra of the carbonyl region of acetic acid and trichloroacetic acid in CCl_4 solution. Their data matrix consisted of the absorbances of 9 solutions of different concentrations measured at 200 and 301 different wavenumbers, respectively. Using their reported eigenvalues we calculated the RE, IE and IND functions which are listed in Table VI.

For acetic acid, only a slight increase in IE (0.00048 to 0.00045) occurs on going from 4 to 5 factors. The IND has a minimum at n = 4. Hence, four factors are present.

Unfortunately, for trichloroacetic acid the IE function exhibits neither a minimum nor a leveling off. No conclusions can be reached on this basis. Fortunately, however, the IND gives

Table VI - Results of Factor Analyzing Digitized Infrared Spectra of the Carbonyl Region of Acetic Acids in $CC\ell_4$ Solutions

	Acetic Acid[a]			Trichloroacetic Acid[b]		
n	RE	IE	IND x 10^5	RE	IE	IND x 10^5
1	0.01461	0.00487	22.83	0.02992	0.00997	46.74
2	0.00284	0.00134	5.80	0.00520	0.00245	10.61
3	0.00174	0.00100	4.82	0.00296	0.00171	8.21
4	0.00072	0.00048	2.90	0.00123	0.00082	4.92
5	0.00060	0.00045	3.75	0.00086	0.00064	5.39
6	0.00046	0.00038	5.13	0.00060	0.00049	6.64
7	0.00036	0.00032	9.12	0.00045	0.00039	11.15
8	0.00024	0.00023	23.90	0.00029	0.00028	29.40

a) Based on a study made by Bulmer and Shurvell, J. Phys. Chem., 77, 256 (1973). Data matrix consisted of the absorbances of 9 solutions of different concentrations measured at 200 different wavenumbers. The error was estimated to be between 0.0005 to 0.0015 absorbance units.

b) Based on a study made by Bulmer and Shurvell, Canad. J. Chem., 53, 1251 (1975). Data matrix consisted of the absorbances of 9 solutions of different concentrations measured at 301 different wavenumbers. The error was estimated to be between 0.0005 to 0.0015 absorbance units.

evidence that four factors are present since it reaches a minimum at n = 4.

The real errors for n = 4, calculated to be 0.00072 and 0.00123, respectively, are within the error range which was estimated to be between 0.0005 and 0.0015 absorbance units. Furthermore, the present conclusion that four factors are responsible for the data is in complete accord with the conclusions of Bulmer and Shurvell based upon other statistical criteria dependent upon a knowledge of the error.

C. Mass Spectra. Ritter and co-workers (4) factor analyzed the mass spectral intensities of various mixtures of the same components. Their purpose was to use FA to deduce the number of components in the related mixtures. Their criteria depended upon a knowledge of the experimental error.

When their data matrix, which consisted of the intensities of 4 different mixtures of cyclohexane and cyclohexene measured at 20 m/e positions, was subjected to FA we obtained the results as shown in Table VII. The IE shows only a very slight decrease on going from 2 to 3 factors. The IND is a minimum at n = 2. Our conclusion that there are two factors agrees with the known fact that only two components are present.

When the mass spectral data concerning 7 different mixtures of cyclohexane and hexane were analyzed the results given in Table VII were obtained. In this case both the IE and IND functions clearly indicated that not two but three components are present. Ritter and co-workers suggested that the unexpected third factor was due to nitrogen contamination. When the intensities of the m/e 28 peaks were removed from the data matrix the IE and IND functions, as shown in Table VII, showed that only two factors remained. This confirmed the suspicion that nitrogen was present.

This also illustrates the sensitivity of the IE and IND functions. The deletion of the m/e 28 data involved the removal of only 7 data points out of 126. The amount of nitrogen present was extremely small and only contributed partially to the m/e 28 intensities.

The RE calculated for the three data matrices were 0.154, 0.128 and 0.134, respectively. These errors are almost three times the error, 0.05, reported by Ritter and co-workers. The error was based solely upon the error involved in reading the intensities from the spectra. The real error from FA is a composite from all sources of error, including operational and instrumental variations as well as reading errors.

D. Gas-Liquid Chromatography. Selzer and Howery (9) factor analyzed the gas-liquid chromatographic retention indices of ethers and found that six abstract factors reproduced the data satisfactorily. Using a data matrix concerning some 22 ethers

Table VII - Results of Factor Analyzing Mass Spectral Intensities of a Series of Related Mixtures[a]

	cyclohexane/cyclohexene[b] mixtures			cyclohexane/hexane[c]			cyclohexane/hexane without m/e 28[d]		
n	RE	IE	IND x 10^2	RE	IE	IND x 10^3	RE	IE	IND x 10^3
1	1.932	0.9660	21.47	1.810	0.684	50.27	1.812	0.685	50.35
2	0.154	0.1092	3.86	0.465	0.249	18.62	0.134	0.071	5.36
3	0.118	0.0969	11.18	0.128	0.084	8.03	0.106	0.070	6.65
4				0.111	0.084	12.30	0.092	0.070	10.25
5				0.098	0.073	24.56	0.072	0.061	18.08
6				0.074	0.068	73.51	0.058	0.054	58.18

a) Data taken from the work of Ritter, Isenhour and Wilkins, Anal. Chem., 48, 591 (1976).
b) Data matrix consisted of the intensities of 4 different mixtures of cyclohexane and cyclohexene measured at 20 m/e position.
c) Data matrix consisted of the intensities of 7 different mixtures of cyclohexane and hexane measured at 18 m/e positions.
d) The intensities of m/e 28 were deleted from the matrix described in c) leaving only 17 m/e positions.

Table VIII - Results of Factor Analyzing the GLC Retention Indices of 22 Ethers on 18 Chromatographic Columns.[a,b]

n	RE	IE	IND	n	RE	IE	IND
1	22.28	5.25	0.07708	10	1.40	1.04	0.02187
2	7.25	2.42	0.02831	11	1.25	0.98	0.02553
3	5.30	2.16	0.02354	12	1.07	0.87	0.02975
4	4.06	1.91	0.02070	13	0.94	0.80	0.03748
5	3.24	1.71	0.01915	14	0.73	0.65	0.04586
6	2.76	1.59	0.01914	15	0.69	0.63	0.07618
7	2.42	1.51	0.01997	16	0.61	0.58	0.15261
8	2.05	1.36	0.02045	17	0.59	0.57	0.59012
9	1.71	1.21	0.02114				

a) This problem was suggested by D. G. Howery, private communication.

b) Data was taken from W. O. McReynolds, "Gas Chromatographic Retention Data," Preston Technical Abstracts Co., Niles, Ill., (1966). The experimental error was estimated to be no greater than 3 retention indices.

on 18 chromatographic columns we obtained the results presented in Table VIII. No conclusions can be drawn from the IE function because IE decreases continuously throughout the entire range of factors, exhibiting neither a minimum nor a leveling off. The IND function, on the other hand, shows a single, true minimum at n = 6 in accord with the conclusion of Selzer and Howery. The RE for six factors is 2.76, in excellent agreement with the experimental error estimation that the uncertainty is no greater than 3 retention indices.

E. Drug Activity. Weiner and Weiner (5) were the first to use factor analysis to investigate biological drug activity. Using the drug data of Keasling and Moffett (10), they factor analyzed the natural logarithm of 11 different responses of 16 structurally related drugs. Because of the nature of the tests no error estimation was reported. Using an arbitrary error criterion Weiner and Weiner found 8 factors.

Using the same data and factor analyzing the natural logarithmic function, we obtained the results shown in Table IX. Notice here that the IE increases as n goes from 1 to 3. Also notice that the IND function blows up as we employ more eigenvectors. This behavior is identical to that which we observed during our analysis of a perfectly random matrix of numbers (see Model Data C in Table IV). We conclude, therefore, that this drug data is not factor analyzable.

There are two possible reasons for this failure. First, the experimental error may be too large. Secondly, it is possible that the logarithm of the drug activity does not obey the sum of product functions demanded by mathematics involved in the FA approach. The acquiring of accurate experimental drug activity data should help eliminate one of these two possibilities. The results presented here, however, warn the analyst that some-

Table IX - Results of Factor Analyzing the Natural Logarithm of the Biological Drug Activity of a Series of Structurally Relative Compounds.[a,b]

n	RE	IE	IND	n	RE	IE	IND
1	0.453	0.136	0.00453	6	0.191	0.141	0.00766
2	0.370	0.158	0.00456	7	0.155	0.124	0.00970
3	0.311	0.162	0.00486	8	0.134	0.114	0.01488
4	0.266	0.160	0.00543	9	0.089	0.080	0.02225
5	0.226	0.152	0.00627	10	0.085	0.081	0.08462

a) The original FA was carried out by Weiner and Weiner, J. Med. Chem., 16, 665 (1973).

b) The drug data was obtained by Keasling and Moffett, J. Med. Chem., 14, 1106 (1971).

thing is wrong. To search for the true controlling factors via target-transformation FA is risky and could be a waste of valuable time.

Remarks

It is important for us to realize that the work described in the latter part of this paper represents a bold attempt to deduce both the dimensions of the factor space as well as the experimental error strictly from a knowledge of the experimental data. The methodology is so new at the present time that its limitations have not been established. Obviously there will exist many situations wherein the IE and IND functions will give misleading results or will fail completely. However, systematic accumulation and documentation of both the successes and the failures of these criteria should eventually lead us to a better understanding of the utility of these functions.

Acknowledgment

The author wishes to express his thanks to Harry Rozyn and John Petchul for help in carrying out many of the calculations.

Literature Cited

1. Funke, P.T., Malinowski, E.R., Martire, D.E., and Pollara, L.Z., Separation Sci., 1, 661 (1966).
2. Weiner, P.H., Malinowski, E.R., and Levinstone, A.R., J. Phys. Chem., 74, 4537 (1970).
3. Hugus, Z.Z., Jr., and El-Awady, A.A., J. Phys. Chem., 75, 2954 (1971).
4. Ritter, G.L., Lowry, S.R., Isenhour, T.L., and Wilkins, C.L., Anal. Chem., 48, 591 (1976)
5. Weiner, M.W. and Weiner, P.H., J. Med. Chem., 16, 665 (1973).
6. Duewer, D.L., Kowalski, B.R. and Fasching, J.L., Anal. Chem. 48, 2002 (1976).
7. Abraham, R.J., Wileman, D.F., and Bedford, G.R., J.C.S. Perkin II, 1027 (1973).
8. Bulmer, J.T. and Shurvell, H.F., J. Phys. Chem., 77, 256 (1973); Canad. J. Chem., 53, 1251 (1975).
9. Selzer, R.B. adn Howery, D.G., J. Chromatography, 115, 665 (1973).
10. Keasling, H.H. and Moffett, R.B., J. Med. Chem., 14, 1106 (1971).

4

The Unique Role of Target-Transformation Factor Analysis in the Chemometric Revolution

DARRYL G. HOWERY

Department of Chemistry, City University of New York, Brooklyn College, Brooklyn, NY 11210

A mathematical-analysis revolution of major import has been occuring in chemistry during the past decade. This symposium is a logical manifestation of the revolution. By adapting a battery of mathematical/statistical techniques to high speed computers, researchers in chemometrics can extract new and heretofore unobtainable insights into large, multifactor data sets. Factor analysis, a major weapon of the revolution, is proving to be a versatile, general method for analyzing matrices of chemical data. In particular, the target-transformation method of factor analysis (1,2), which enables one to test empirical and theoretical models, offers powerful and unique potentialities for obtaining partial and even complete solutions to many kinds of chemical problems. The main objective of this presentation is to summarize the distinctive attributes of target-transformation factor analysis (TTFA).

Factor analytical solutions are of a form nicely adapted to chemistry. A data point, d_{ij}, in a data matrix is expressed as a linear sum of factors, each factor being the product of a row-designee cofactor and a column-designee cofactor. Mathematically, factor analytical solutions obey the equation:

$$d_{ij} = \sum_{m=1}^{n} d_{ij,m} = \sum_{m=1}^{n} r_{im}\, c_{mj} \qquad (1)$$

where n is the minimum number of factor terms, m, to adequately predict the data, and r_{im} and c_{mj} are the cofactors for the $\underline{j}^{th}$ row designee and the $\underline{i}$th column designee, respectively, associated with the $\underline{m}$th factor. The central purpose of a TTFA is to derive information about the two sets of cofactors not only in an abstract (mathematical) sense but also in a real (physically significant) sense. In matrix notation, the data matrix

$$[\,D\,] = [\,R\,]\,[\,C\,] \qquad (2)$$

where [R] is the row matrix containing a row for each row

designee and a column for each cofactor, and [C] is the column matrix having a column for each column designee and a row for each cofactor.

Solutions of the type indicated by equations (1) and (2) are especially suited to studies of entity-entity data matrices. The abstract factors are related in some manner to those real factor terms which measurably influence the data. Even more conveniently, each cofactor pair in a given factor can be transformed via target transformation to specific properties of the row designees and the column designees which are responsible for the factor. In a solute-solvent problem, for example, the factors correspond to the important energies of interaction and the cofactors pinpoint the nature of the interactions in terms of the properties of the pure solute and the pure solvent. (In usual chemical particles such as molecules, ions and radicals, but also, e.g., biological species, persons, political groups and celestial bodies.)

Factor analytical research can be based upon the target-transformation technique and/or upon what we term the abstract factor analytical approach. In abstract (traditional) FA, abstract solutions are obtained under various mathematical constraints. The investigator then tries to gain insight by examining the coefficients in the matrices generated in the abstract solution. Abstract FA, long used in the psychological and social sciences, can be gainfully applied to certain types of chemical problems (3,4). The target-transformation method of Malinowski (1) opens new territory by enabling the researcher to test parameters of the row and column designees of the matrix. The TTFA extension, in allowing one the possibility of transforming from abstract factors to real factors of the designees, alleviates for the physical scientist a major weakness of abstract FA.

The steps in a complete TTFA: data preparation, reproduction, target transformation, combination and prediction, have been discussed elsewhere (5). The number of factors required in equation (1) can be estimated in the short-circuit reproduction procedure using both experimental-error and theoretical criteria, as was deftly explained in the previous talk (6). The model-testing capability of the target transformation step is the heart of factor analysis for the physical scientist. A best complete model, i.e., the best real solution, is generated in the combination step.

TTFA has been thoroughly tested during the past six years. Howery (5) and Weiner (7), in recent reviews which complement each other, consider the philosophy, theory, procedures and applications of TTFA. Details of the mathematical development are given in the already classic paper of Malinowski and coworkers (2). TTFA can be utilized using a blend of theroretical and empirical insights. Important TTFA's based at least in part on a theoretical framework include the study of solute-solvent interactions influencing proton chemical shifts (2,8), the verification of gas chromatographic retention mechanisms (9), and the elucidation of solute-solvent effects on acidity constants (10). These papers illustrate the exceptional power of TTFA if the factor analyst starts with some theoretical help. In such cases, in-depth fundamental solutions can be achieved. However, for most chemical problems, theoretical insight is minimal. Thus, the second way of using TTFA, involving a more empirical approach, has an even wider applicability in chemistry. Examples of empirical solutions include a detailed study of ether cofactors (11) (one of a series of researches on the solute cofactors influencing retention indices), and an investigation of solvent-metal effects on polarographic half-wave potentials (12). These studies show the potential for using TTFA to furnish useful empirical solutions in fields devoid of a theoretical underpinning.

Target Transformations

Unique Features. The quite unique attributes of the target-transformation procedure center around the model-testing and model-building capabilities of TTFA. No other mathematical/statistical method shows such promise for extracting real cofactors and for developing complete solutions to multifactor problems.

1) Potential cofactors are separated and tested mathematically regardless of the complexity of the data space. Any parameter of either the row or column designees can be investigated independently. Single terms in a theoretical or empirical model can be tested one at a time even though the other cofactors in the space are operative, an unmatched accomplishment of TTFA. Target transformation serves to curve fit vectors of parameters in a multifactor space.

2) Restrictions on the procedure are minimal. No knowledge of the other cofactors is required. One can start with complete ignorance of the nature of the real cofactors, in marked contrast with multiple regression analysis which is applicable only if a complete model is specified. Furthermore, the real vectors to be tested need not be complete, a tremendous practical advantage since the data store for most types of chemical information is usually incomplete. Missing or uncertain points can be left blank on a test vector (a procedure termed "free floating"). Such points will be predicted as a premium in successful target transformations.

3) The separation of the factor analytical solution into two parts as shown in equation (2) enables one to build up solutions for the two kinds of designees independently. Even if the problem in terms of one kind of designee appears hopelessly complex, it is still possible to derive a solution for

the other kind of designee. Two complete solutions can be developed term by term and each possible real solution involving sets of cofactors can be tested in the combination step. If all of the factors in the abstract solution are not spanned in a given combination, the reproduction via combination will be poor, indicating the sensitivity and the non-force-fitting nature of the step.

Two Examples. Two typical examples from recent research will illustrate the scope of the target-transformation approach for testing parameters of the designees. Computations were carried out on an I.B.M. 370/168 digital computer using a computer program in FORTRAN IV which has evolved over the decade (13). Vectors to be tested can contain data of any type which the researcher thinks might be indicative of the behavior of the designees (and hence possibly responsible for a cofactor). Both physical vectors (illustrated by example one) and structural vectors (exemplified by the second example) can be tested. Descriptors used in pattern recognition studies have much in common with the test vectors of TTFA. The essential question in evaluating the success of a target transformation is how well does the best-fit predicted vector calculated from the least-squares method (2) agree point-by-point with the real vector being tested. If the two vectors are reasonably similar, the test vector is taken to be a real cofactor. The examples to be sited involve for pedagogical purposes a difficult-to-interpret result and an unsuccessful transformation.

The first example is taken from a TTFA of the retention indices of organic solutes on stationary-phase solvents (14). Studies of retention indices have amply demonstrated the ability of TTFA to isolate cofactors in problems far too complicated for detailed theoretical treatments. Whereas solutes have been studied in detail, this is the first in-depth investigation of GLC solvents using TTFA. To better examine the cofactors of the solvents, only monomeric solvents were selected. (Previous TTFA's of retention indices have involved relatively complex, polymeric solvents for which test vectors are difficult to formulate.) The specific parameter tested by target transformation in this example is the molar refraction, a vector which has generally tested well as a solute cofactor. As shown in Table I, agreement between the test vector and the predicted vector is overall moderately good at best. Values for three deliberately free-floated points are predicted reasonably well. The molar refraction may be a cofactor; such borderline conclusions are common in TTFA.

The second example is selected to illustrate the manner in which structural vectors based on chemical insight can be employed to track down cofactors. Such vectors are especially useful for developing empirical solutions. The example involves bond dissociation energies for radical-radical bonds (15), a

Table I - Target transformation of a physical test vector.

Data matrix: retention indices for 39 carbonyl-containing solutes on 18 monomeric stationary-phase solvents, data taken from reference 16.

Test vector: molar refraction estimated by summing specific refractions, TT in 6-factor space, free-floated points shown in parentheses.

Solvent	Test Vector	Predicted Vector	Solvent	Test Vector	Predicted Vector
bis(2-ethoxyethyl) phthalate	(79.7)	101.0	Hallocomid M18	(140.8)	129.7
dibutyltetrachloro phthalate	96.6	98.6	Hyprose SP80	203.7	163.0
di-2-ethylhexyl adipate	107.7	138.2	isooctyldecyl adipate	119.6	134.1
di-2-ethylhexyl sebacate	126.3	133.8	Quadrol	76.3	117.9
diglycerol	37.7	39.0	sucrose acetate isobutyrate	192.4	176.8
diisodecyl phthalate	134.7	110.3	sucrose octaacetate	141.3	166.2
dioctyl phthalate	113.6	105.9	TMP tripelargonate	160.5	154.5
dioctyl sebacate	126.3	132.0	tricresyl phosphate	103.0	84.5
Flexol 8N8	139.3	125.4	Zonyl E7	(134.8)	117.6

Table II - Target transformation of a structural test vector.

Data Matrix: bond dissociation energies involving 12 radicals with the same 12 radicals, data taken from compilation of A. Zavitsas (17).

Test vector: halogen uniqueness, TT in 3-factor space.

Radical	Test Vector	Predicted Vector	Radical	Test Vector	Predicted Vector
methyl	0.00	0.02	fluorine	1.00	0.97
ethyl	.00	.04	chlorine	1.00	.34
isopropyl	.00	.06	bromine	1.00	.17
t-butyl	.00	.06	hydroxy	0.00	.27
phenyl	.00	.08	methoxy	.00	.28
benzyl	.00	.04	amine	.00	.47

most basic type of chemical data. The test vector shown in Table II is designed to ascertain if the group of halogen-radicals (assigned test values of "1") is responsible for a unique cofactor, i.e., a cofactor not exhibited by the remaining group of radicals (given test values of "0"). Such cofactors can be tested without knowing the theoretical form of the interaction term. As can be seen in Table II, the target transformation is clearly unsuccessful; a unique halogen interaction is not a cofactor.

Ramifications of TTFA

Combination. Sets of vectors can be utilized simultaneously in the combination step to find the best empirical solution. This step of FA is similar to multiple regression analysis in that complete models are utilized, but different from regression analysis in that FA does not prescribe that the model be force fitted. If any factor is missing in a proposed model, the sensitive combination step will lead to a poor reproduction. Even so, solutions having errors less than twice experimental error have been developed from thorough factor analyses of retention indices using the empirical approach. For example, in the problem referred to in the first TT example above, the best solution to the solvent part of the complex problem gave an error of 7.1 r. i. units; the empirical model can predict r. i.'s with an error of about one percent.

Prediction. The predictive ability of TTFA has as yet received little attention. To illustrate the potential of this step, consider the prediction of a new row of data based on the best empirical solution obtained in the combination step. To calculate a new data point associated with an added row designee, x, and a column designee from the original data matrix, j, a modified form of equation (1) is employed:

$$d_{xj} = \sum_{m=1}^{n} r_{real,\, xm} c_{calc,mj} \qquad (3)$$

The row-designee cofactors in equation (3), r_{real} are those key vectors from the best solution via combination, while the column-designee cofactors, c_{calc}, are coefficients in the [C] matrix (which is readily calculated using equation (2), given a solution [R_{real}] and the data matrix). To calculate the new datum, only the values of the $\underline{n}$ real cofactors for the new designee and the $\underline{n}$ coefficients in the $\underline{j}$th row of the calculated column matrix are required. For example, in the study of the cofactors of ethers (11), a real solution having the following six vectors: carbon number, total atom number, chain difference, chain ratio, total atom number squared, and boiling point squared, produced an error of only 5.4 r.i. units. The retention index of t-butylmethyl ether (a solute not incorporated in the original matrix) on butyltetrachloro phthalate (a solvent in the original matrix) is predicted using equation (3) to be 613 r. i. units, nicely in agreement with the experimental value of 609 ± 3 units. The mean error for the entire new row involving the new ether with the 25 original solvents is 3.5 units. Such satisfactory predictions indicate that the empirical solution quite adequately spans all of the solute cofactors.

The examples put forth during this presentation demonstrate clearly that the TTFA approach can be utilized to thoroughly characterize a chemical data space. Target transformation methodology seems destined to play a leading and unique role in the chemometric revolution.

Literature Cited

1. Malinowski, E. R., Doctoral Dissertation, Stevens Inst. Technology, Hoboken, N. J., 1961.
2. Weiner, P. H., Malinowski, E. R., and Levinstone, A., J. Phys. Chem., (1970), 74, 4537.
3. Bulmer, J. T., and Shurvel, H. F., J. Phys. Chem., (1973), 77, 256.
4. Rozett, R. W., and Petersen, E. M., Anal. Chem., (1976), 48, 817.
5. Howery, D. G., Amer. Lab., (1976), 8(2), 14.
6. Malinowski, E. R., in "Chemometrics: Theory and Applications," B. R. Kowalski, Ed., A. C. S. Symposium Series, P. xxx, 1977.
7. Weiner, P. H., Chem. Tech., in press.
8. Weiner, P. H., and Malinowski, E. R., J. Phys. Chem., (1971), 75, 3160.
9. Weiner, P. H., Liao, H. L., and Karger, B. L., Anal. Chem., (1974), 46, 2182.
10. Weiner, P. H., J. Amer. Chem. Soc., (1973), 95, 5845.
11. Selzer, R. B., and Howery, D. G., J. Chromatogr., (1975), 115, 139.
12. Howery, D. G., Bull. Chem. Soc. Japan, (1972), 45, 2643.
13. Malinowski, E. R., Howery, D. G., Weiner, P. H., Soroka, J. M., Funke, R. T., Selzer, R. B., and Levinstone, A., "FACTANAL - Target-Transformation Factor Analysis," Program 320, Quant. Chem. Prog. Exch., Indiana Univ., Bloomington, Ind., 1976.
14. Soroka, J. M., and Howery, D. G., to be submitted.
15. Howery, D. G., to be submitted.
16. McReynolds, W. O., "Gas Chromatographic Retention Data," Preston Tech. Abstracts Co., Niles, Ill., 1966.
17. Zavitsas, A., Long Island Univ., Brooklyn, N. Y., private communication.

5

Application of Factor Analysis to the Study of Rain Chemistry in the Puget Sound Region

ERIC J. KNUDSON, DAVID L. DUEWER, and GARY D. CHRISTIAN

Department of Chemistry, University of Washington, Seattle, WA 98195

TIMOTHY V. LARSON

Department of Atmospheric Sciences, University of Washington, Seattle, WA 98195

As part of a study on "acid rain" at the University of Washington, we have investigated the applications of factor analysis to understanding the variations in chemical compositions of rain samples gathered during single storms in the Puget Sound region. The determination of the sources of such variation, and the extent of their influence, can aid not only in understanding air chemistry in general, but also in making future judgements concerning man's activities which affect atmospheric chemistry.

The collection of rain samples has several points to recommend it as an ideal means for sampling the chemical state of the atmosphere. It is relatively inexpensive and simple to do. Much of the aerosol content of air, as well as many gaseous constituents, are apparently removed by precipitation (1). Rain consists primarily of distilled water in which a relatively uniform background spectrum of chemical constituents (mainly sea salts) is dissolved. And local variations in the chemical composition of rain samples are primarily a function of local variations in atmospheric chemistry due to local sources, as mediated by the rates of rainout and washout and variations in rain volumn.

The data used in this paper are from analyses of rain samples gathered over a 24 hour period during a single storm in 1975. The factor analysis techniques used are generally widely known and are available, at least in part, in a number of computer data analysis systems. The system used in this study, ARTHUR, is available from the Laboratory for Chemometrics at the University of Washington (2).

Experimental

Rain samples were collected in washed polyethylene buckets placed at pre-selected sites in the Puget Sound region (see map, Figure 1). After a sampling period of 24 hours during a single storm, the samples were returned to the laboratory and transferred to washed polyethylene bottles, from which aliquots were taken for analysis. Species analyzed for were sodium, potassium, calcium, magnesium, zinc, copper, cadmium, manganese, lead, arsenic, and antimony, as well as hydrogen, ammonium, nitrate, sulfate, and chloride ions. Details of analytical procedures used (atomic absorption spectrometry for the metals and metalloids, and other methods for the other ions) are discussed elsewhere (3,4).

Factor analysis techniques used were the following: derivation of the correlation matrix and a heirarchical dendrogram based on it, eigenvector and varimax-rotated vector representations, and the derivation of "features" from the eigenvector and varimax-vector representations via the Karhunen-Loeve transform (and an analogous transform utilizing the varimax-rotated vectors). Analytical error perturbations on the original data set were used to evaluate the stability of the eigenvector and varimax-rotated vector representations. Creation of contour maps of single species and of multivariable ("feature") representations was accomplished through the use of the SYMAP package of computer programs developed at the Laboratory for Computer Graphics of the Harvard University Graduate School of Design.

Results and Discussion

Single Species Maps. A summary of the concentrations of the species found in the rain samples is given in Table I, and representative contour maps of several of these species are given in Figures 2 through 10, plotted over the geographic sampling area. The predominant movement of this storm, as that of most storms through the Puget Sound region, was from generally southwest to northeast. Hence, higher concentrations of a species in a given area is likely to be indicative of a source for that species lying to the southwest of the elevated concentrations.

It can be seen, for example, that there appears to be a source of both arsenic and antimony (Figures 2 and 3) in or near Tacoma. In this case, the source, well known to residents of the region, is a large copper smelter located in Tacoma. The effects of this source are also apparent in copper (Figure 4) and, to a certain extent, cadmium (Figure 5) maps.

Hydrogen ion concentrations (Figure 6) are elevated downwind of both Seattle and Tacoma, a not unexpected phenomenon, but sulfate ion concentrations (Figure 7) seem indicative of no particular source. This is somewhat contrary to observations on two earlier storms sampled (and of several other reported studies on "acid rain"), where hydrogen and sulfate ion concentrations were found to be strongly correlated. In this particular storm, however, it is nitrate ion concentrations (Figure 8) which appear to be associated with increased hydrogen ion concentrations.

The contour map for zinc concentrations (Figure 9) shows no identifiable anthropogenic influence, while the contour map for lead (Figure 10) illustrates the problem encountered when the "plume" lengths are considerably shorter than the sampling distances. Lead concentrations are undoubtedly associated with traffic patterns, and the map is dominated by "hot spots," probably indicative of how close the sampling bucket was placed to areas of heavy traffic.

Correlation Matrix and Heirarchical Dendrogram. The correlation matrix is readily derived from the normalized data matrix, D_n, by multiplying the transpose of this matrix by the matrix D_n

$$C = D_n^T \cdot D_n$$

The correlation matrix for the analytical data presented in Table I is given in Figure 11: the upper half is the complete correlation matrix, and the lower half consists of only those correlations greater than or equal to 0.7.

Examination of this matrix shows some obvious groupings of highly-correlated species. The largest group of highly intercorrelated species contains sodium, potassium, calcium, magnesium, and chloride ion. Another prominent group contains arsenic, antimony, cadmium, and copper. And finally, there is a small group consisting of hydrogen and nitrate ions. Another striking feature is the non-correlation of hydrogen and sulfate ion concentrations, as mentioned earlier.

A simplified way of looking at the correlation matrix is a heirarchical "similarity" dendrogram. The technique of deriving this representation involves converting the correlations to "distance" representations by subtracting the absolute value of the correlation coefficients from 1.0, then subjecting the resulting matrix to hierarchical Q-mode clustering, as described by Kowalski and Bender (5). The resulting dendrogram is given in Figure 12.

In this dendrogram (as suggested by the correlation matrix), three main groups predominate. The largest group consists of sodium, magnesium, chloride, calcium, potassium, and ammonium ion: these species are probably associated with a sea-salt background. The second group consists of those species presumably emitted by the copper smelter mentioned earlier: copper, arsenic, antimony, and cadmium. And the third grouping consists of hydrogen and nitrate ions. There remain only single-element "groups," showing little relationship to the other species: zinc, manganese, lead, and sulfate. The singularity of lead is relatively easy to explain on the basis of many small point sources and short "plumes" mentioned earlier. Zinc may be split between association with smelter-emitted elements and with sea salts, resulting in relatively low correlations to either group. The singularity of manganese and, especially, sulfate are difficult to explain. In the case of sulfate, it may be possible that reduced emissions of sulfur dioxide from the smelter (from levels noted in earlier samplings, where there were much higher correlations between hydrogen

Table I. Raw Data[a]

STATION	VOL.	PH	SO4	NH4+	CL-	NO3-	NA	K	CA	MG	ZN	CD	AS	SB	CU	PB	MN
1 U OF WASH.	57	4.35	3700	315	2680	460	2000	140	550	190	23	0.45	1.8	0.93	29	54	5.7
2 SO.TAC.APT	51	4.75	3000	240	180	355	290	170	200	55	15	0.28	0.5	0.05	15	20	7.1
3 DASH POINT	66	4.00	2600	20	1680	900	1800	210	250	160	6	0.25	3.8	0.46	40	10	4.0
4 STEILACOOM	82	4.75	1500	110	230	300	230	240	105	50	6	0.67	0.6	0.08	13	3.0	4.7
5 AUBURN	104	4.60	1300	525	4090	430	2300	140	250	240	14	0.16	1.0	0.12	6	15	2.5
6 ZENITH	85	4.05	3000	340	2820	630	2600	650	730	220	[illegible]	0.58	8.0	1.04	67	16	6.0
7 EVANS PT.	62	4.20	1300	96	820	525	800	170	125	115	13	0.30	1.0	0.14	21	5	2.1
8 3-TREE PT.	75	3.95	1300	82	770	1250	1100	120	120	40	9	0.30	4.0	0.15	25	[illegible]	2.1
9 RAINIER B.	270	4.45	1600	104	680	215	550	35	75	55	4.5	0.18	16.0	0.80	60	12	1.0
10 LINCOLN PK	240	4.65	2500	60	550	170	540	150	100	50	7	0.15	1.0	0.15	10	7	1.6
11 ALKI POINT	150	4.40	1400	54	1270	180	1300	140	210	120	11	0.35	1.0	0.20	32	9	1.5
12 WASH. CTR.	69	4.45	4300	110	1590	435	1800	190	235	195	12	0.29	4.0	0.73	21	7	2.2
13 TALLEQUAH	30	4.45	4300	116	1590	435	1800	190	235	195	12	0.29	4.0	0.73	53	7	4.0
14 ROBINSON PT	59	4.00	2500	186	2820	630	2600	240	300	165	22	0.92	47	2.06	290	21	4.2
15 CEDAR MT.	277	4.45	2200	114	540	240	450	40	90	75	1.5	0.08	5.0	0.46	9	5	1.2
16 YELLOW LK.	298	4.50	3100	50	770	305	850	55	72	95	25	0.09	7.7	0.28	110	6	1.3
17 KENILWORTH	142	4.15	4300	30	2640	1125	2400	500	190	270	5	0.17	4.5	0.40	51	56	2.3
18 DUVALL	142	4.30	2700	280	2180	465	2100	320	290	250	17	0.36	4.0	0.38	60	10	4.8
19 CLEARVIEW	119	4.05	2200	114	500	385	900	110	150	125	11	0.07	5.5	0.05	8	11	3.4
20 NORMA BCH.	132	4.35	1100	24	820	505	1100	240	110	125	13	0.36	0.7	0.03	11	3.5	4.0
21 JUANITA	132	4.35	2200	106	1360	500	1700	90	230	170	6	0.09	7.0	0.30	21	23	2.9
22 REITER	230	4.50	3500	60	360	410	360	33	75	45	2.5	0.02	1.5	0.12	2	3.0	1.3

[a] Values given as ppb of species listed. Volumes in milliliters per bucket. Refer to Figure 1 for station locations.

and sulfate ions) tended to give greater importance to other sources of sulfate concentrations in the sampling area.

Eigenanalysis. Diagonalization of the correlation matrix, C, results in an orthonormal set of eignevectors, E_j, with their associated eigenvalues, j (where j = 1 to n, the dimension of the correlation matrix). Rearrangement of these eigenvectors in order of decreasing magnitude of their eigenvalues reflects the ordering of the amount of variance in the original data set spanned by the individual eigenvectors: the quotient of the eigenvalue (for a given eigenvector) divided by the sum of all the eigenvalues reflects the relative amount of "information" about the data matrix (variance) contained in that eigenvector.[5,6]

The results of diagonalization and rearrangement of the 1975 data matrix are given in Table II and are summarized graphically in Figure 13. These histograms are drawn as follows: each histogram represents one eigenvector and is linearly scaled to the ratio of its associated eigenvalue to the eigenvalue for the first eigenvector. Within each eigenvector histogram, the "percent information" represents the coefficient ("loading") of each species' concentration divided by the sum of the coefficients in that vector (times 100). Prior to diagonalization, concentrations of all species are normalized to a mean of zero and a standard deviation of 1.0, and the coefficients in the vectors (and histograms) are coefficients of the normalized concentrations.

The first eigenvector consists of significant contributions from all determined species, ranging from 12.3% information from sodium to 2.6% information from nitrate. While the largest contributions are from elements typically associated with "sea salts," contributions from other species are by no means minor. In fact, this particular vector is most highly correlated to the total cationic charge concentration. This particular vector spans 42.3% of the variance in the data set (see Table II), indicating that the largest principal component of this data set consists of total dissolved ion concentrations.

The second eigenvector, spanning 15.2% of the variance, consists mainly of elements probably emitted from the smelter in Tacoma: arsenic, copper, antimony, and cadmium. This vector also contains significant contributions from other species, probably a result of the constraints that a) the second eigenvector is perpendicular to the first, and b) the second eigenvector contains the maximum remaining variance. The third eigenvector spans 11.6% of the variance and is predominantly nitrate and hydrogen ion concentrations, with, again, some contributions from other species.

That these three eigenvectors are the largest principal components can be more or less predicted by the three distinct groupings of intercorrelated species in the correlation matrix, for which interpretations have already been given. Together, these three eigenvectors account for almost 70% of the variance in the data.

Table II. Eigenvectors[a]

EIGENVALUE	VAR.	H+	NH4+	NA	K	CA	MG	ZN	CU	PB	SO4	MN	CD	AS	SB	CL-	NO3-
1 6.76900	42.3	.194	.148	.351	.258	.310	.311	.238	.236	.225	.162	.205	.239	.204	.267	.321	.161
2 2.42600	15.2	-.182	.253	.092	.165	.197	.227	.063	-.473	.205	.106	.164	-.255	-.506	-.345	.163	.034
3 1.85900	11.6	.464	-.375	.130	.132	-.153	.145	-.277	[illegible]	[illegible]	[illegible]	-.263	-.226	-.042	-.099	-.002	[illegible]
4 1.30900	8.2	.261	-.035	-.102	.347	.150	-.143	-.016	-.130	-.328	-.408	.407	.418	-.149	-.151	-.205	.196
5 1.15000	7.2	.080	.368	.212	-.133	-.053	.176	-.082	-.023	-.277	-.622	-.375	-.061	.071	-.074	.360	-.011
6 .67660	4.2	-.205	-.010	.089	.430	.019	.137	-.707	-.004	-.246	.227	-.021	.058	.100	.131	.080	-.263
7 .46680	2.9	-.253	-.366	.166	.330	-.275	.297	.458	.233	-.279	.014	-.134	.048	-.115	-.283	.040	-.085
8 .44740	2.8	.362	.129	.064	-.161	.192	.054	.174	.035	-.652	.353	.145	-.400	-.030	.004	-.061	-.075
9 .35650	2.2	.142	-.263	.003	.253	.570	-.189	.162	-.129	.066	-.116	-.504	-.067	-.159	.213	-.075	-.304
10 .22010	1.4	-.133	.468	-.298	.369	-.062	-.364	.143	.146	-.107	.231	-.345	-.017	.014	-.053	.033	.373
11 .16440	1.0	.240	.156	-.146	.346	-.117	.113	.047	.101	.272	-.213	.150	-.443	.393	-.220	-.251	-.312
12 .08173	.5	.509	.136	-.172	-.047	-.232	.030	-.019	.006	.108	.261	-.204	.447	-.193	-.264	.148	-.397
13 .03505	.2	-.013	.160	-.250	.034	-.192	.541	.145	-.366	-.064	.004	-.160	.116	-.018	.464	-.402	.075
14 .02757	.2	-.145	.129	.194	-.198	.358	.204	-.106	.095	.023	.107	-.230	.235	.253	-.476	-.500	.114
15 .00609	.1	-.006	.084	-.246	-.060	.135	.246	-.195	.671	.084	-.157	.032	-.112	-.533	.109	-.140	.023
16 .00249	0.0	.070	.236	.676	.083	-.320	-.283	-.002	.048	.030	-.006	-.028	.001	-.286	.173	-.402	-.103

[a] Eigenvectors are arranged vertically by relative magnitudes of their eigenvalues. Columns are loadings on species listed at the top. "VAR." refers to percent of total variance in that eigenvector (eigenvalue divided by sum of the eigenvalues).

The remaining eigenvectors probably do not represent anything meaningful in terms of explaining the influences on (or factors in) the observed data. Most of them are most likely due to sampling and/or analytical variations.

Varimax Rotation. A method for "cleaning up" the eignevalue-eigenvector representation of the data is the varimax rotation. This procedure maximizes the variance in each vector, the effect of which is to decrease the number of variables with intermediate loadings and to increase the number of those with large and small loadings in each vector.[7,8] This results in "cleaner" vectors containing most of their information in only a few variables. The varimax-rotated vectors (which will be hereinafter called "varivectors" for want of a better term) are given in Table III and histograms are given in Figure 14.

It can be seen that the varivectors are indeed "cleaner." It can also be seen that some rearrangement of vectors has taken place, in terms of not only the relative amount of variance spanned per vector, but also the ordering of vectors (in terms of their "varivalues," a term which will be used to represent the length of a unit vector on the varivector, analogous to the term "eignevalue"). The magnitude of the first two varivectors (sea-salts and smelter elements), is approximately equal, in contrast to the eignevectors, where the first eigenvector spans approximately three times the variance of the second. But, as mentioned earlier, the first eigenvector includes contributions from all species and is that vector spanning the maximum amount of variance over the entire data set, while the corresponding second varivector includes large contributions from three species and small contributions from nine species (and it is not constrained to include the maximum variance over the data set).

Similarly, the varivector spanning the greatest variance contains major contributions from only four species: arsenic, copper, antimony, and cadmium. And the third varivector consists almost exclusively of nitric acid. Almost all of the remaining varivectors consist of single variable loadings, leading to the assumption that the data set consists of three major principal components and several single-species factors. Figure 15 illustrates this point: the first fifteen eigenvectors are given in order (columns), and the varivectors are rearranged to correspond to the best fit with the eigenvector order. In this figure, the area of each box corresponds to the square of the loading (coefficient) for that variable in that particular vector.

A series of data were obtained by perturbing the original data with assumed analytical errors. These assumed errors are listed in Table IV, and are given as one standard deviation (relative, per cent).

Table III. Varimax-rotated Vectors (rows and columns as defined for Table II)[a]

	VARIVALUE	VAR.	H+	NH4+	NA	K	CA	MG	ZN	CU	PB	SO4	MN	CD	AS	SB	CL-	NO3-
1	3.26100	20.4	-.173	-.011	-.163	-.033	-.004	-.045	-.135	-.018	-.027	-.064	-.013	-.323	-.545	-.479	-.119	.004
2	3.13000	19.6	.096	.237	.479	.218	.241	.512	.130	[illegible]	[illegible]	.047	.068	.046	.053	.124	.477	.130
3	1.51700	11.4	.637	-.124	.194	.174	.070	.140	-.006	.077	.093	.011	.021	.043	.070	.051	.073	.676
4	1.43100	8.9	.032	.185	.091	.232	.294	.105	.181	.009	.067	.092	.776	.389	-.008	.079	.004	.012
5	1.13600	7.1	.040	.056	-.097	-.101	-.120	-.146	-.048	-.092	-.252	-.653	-.129	.192	.015	-.160	-.048	-.059
6	1.05900	6.6	-.012	-.145	-.135	-.024	-.240	-.104	-.865	-.224	-.136	-.044	-.173	-.158	-.004	-.067	-.120	.009
7	1.01500	6.3	-.073	.033	.093	.025	.247	.097	.135	-.048	.067	-.038	.161	.053	.035	.082	.355	-.068
8	.98730	6.2	.074	.019	.160	.330	.289	.184	.027	.057	.060	.075	.177	.291	-.020	.035	.103	.142
9	.98550	6.2	.036	-.061	-.130	-.068	-.163	-.152	-.134	.035	-.879	-.213	-.065	-.015	-.019	-.132	-.189	-.166
10	.63030	3.9	.229	.139	.216	.145	.790	.000	.113	-.095	.087	.059	.117	.125	-.046	.379	.116	-.122
11	.30770	1.9	.072	-.019	-.028	-.118	-.100	.072	-.068	-.099	-.010	.063	-.071	-.933	.095	-.161	-.163	-.094
12	.14090	.9	.779	-.004	-.064	.001	.072	.117	.002	-.002	-.017	-.011	.007	-.036	.096	-.049	-.071	-.593
13	.04224	.3	.020	-.006	.152	-.003	.018	-.065	-.007	.383	-.014	-.012	-.012	-.024	.221	-.880	-.123	-.021
14	.03937	.2	.036	.029	-.080	-.003	-.010	.619	-.004	-.174	-.004	-.001	.010	-.025	.156	-.055	-.741	-.032
15	.01009	.1	-.006	-.011	-.090	.006	-.006	.011	.003	.762	-.005	.009	.001	.012	-.634	-.002	.097	.011
16	.00381	0.0	-.011	.011	.845	.002	.008	-.478	.002	-.028	.002	.001	.004	.001	.057	-.027	-.228	.007

[a] "Varivalue" = length of basis vector, analogous to "eigenvalue." "VAR." = percent of total variance contained in that varivector (*see* Table II).

Table IV. Assumed Analytical Errors

Species	Error	Species	Error
H^+	6.7	Pb	2.6
NH_4^+	6.7	$SO_4^=$	20
Na	3.4	Mn	2.6
K	3.4	Cd	2.6
Ca	3.4	As	4.0
Mg	6.7	Sb	4.7
Zn	3.4	Cl^-	10
Cu	2.6	NO_3^-	6.7

Using these error values, a random-number generating program was used to generate eighteen new sets of data from the original data.

From each set of these error-perturbed data, correlation matrices and eigenvector and varivector representations were derived. After matching the derived vectors in each set to the corresponding vectors in the original set (since the perturbations produced some mixing of the eigenvalues and varivalues), means and standard deviations were calculated for the loadings within each vector. These data are also plotted in Figure 15: the vertical and horizontal bars inside the boxes represent the means plus and minus one standard deviation, respectively, of the loadings.

What is demonstrated is that the loadings for at least the first three eigenvectors are more or less invariant to the expected analytical errors. After the first three eigenvectors, uncertainty in the loadings becomes significant. Hence, interpretation based on analysis of the eignevectors is more or less limited to the first three eigenvectors as factors or representations of factors. On the other hand, the identity of the varivectors remains essentially unchanged by the introduction of analytical error (even though some reshuffling of varivalues did take place, necessitating considerable resorting to match vectors with the original set). The reason for this is that the varimax rotation maximizes the variance in each vector into as few variables as possible, and minimizes the loadings on the rest of the variables. And since all of the varivectors after the first two contain high loadings for only one or two variables, the errors in the remaining variables become small by comparision and contribute little to variability in loading in the one or two major variables (in contrast to the situation observed in the set of eigenvectors). Consider, for example, the fifth eigenvector and the fifth varivector: for the non-error-perturbed data, 39% and 78%, respectively, of the variance in these vectors is sulfate concentration. For the eighteen sets of error-perturbed data, the sulfate loadings in the fifth eigenvector varies from 20% to 53% of the total loading, with a mean of 34% and a standard deviation of 13.6%. For the corresponding fifth varivector, the sulfate loadings range from 68% to 89% of the total loadings, with a mean of 79% and a standard deviation of 4.7%. Hence, the constraints on the eigenvector representation (orthogonality, and maximum variance over the data set), and the constraints on the varimax rotation (not necessarily orthogonal, and maximizing the variance within each vector) produce considerably different results when error perturbations are introduced: the former results in considerable "noise" in the loadings of vectors having smaller eigenvalues, while the latter seems to be relatively insensitive to analytical error.

Features. Principal component maps over the sampling area can be constructed via the Karhunen-Loève transform.[9] This transform consists of multiplying the normalized data matrix by the transpose of the eigenvector matrix. The result is a "feature" matrix of dimensions m by n, where m is the number of sites and n is the number of variables determined. Put more simply, the first feature value for the first site is the sum of the products of the normalized observed values at the first site and the corresponding loadings in the first eigenvector. The second feature value for the first site is the sum of the products of the normalized observed values at the first site and the corresponding loadings in the second eignevector. And so on. These features can each then be mapped onto the sampling area to obtain a representation of the goegraphic variation of each feature, by using the same mapping program as previously described. A similar set of features which shall be called "varimax features," (for lack of any other known terminology) may be obtained by multiplying the data matrix by the transpose of the varivector matrix. The results of these operations, the Karhunen-Loève features and the varimax features, are given in Tables V and VI.

Maps of the first three Karhunen-Loève features and the first three varimax features are given in Figures 16 through 21. It can be seen that there is very little qualitative difference between the maps for the first Karhunen-Loève feature and the second varimax feature. This is not surprising, considering the correlation between the first eigenvector and the second varivector (0.966). Both maps can be compared to the volume map, Figure 22, which shows roughly opposite behavior. The correlations between volume and the Karhunen-Loève and varimax features are -.52 and -.49, respectively. This result fits well with the observation by many investigators that the concentrations of most species are negatively correlated to rain volume (see reference 1).

A comparison of the maps for the second Karhunen-Loève feature and the first varimax feature, however, shows some striking differences. Recalling that the vectors which give these features represent primarily smelter elements, it might at first glance appear that the second Karhunen-Loève feature is a better picture of the smelter "plume." But two things should be considered: first, the wind was not blowing in the direction indicated on the

Table V. Karhunen–Loève Features[a]

STATION	KL1	KL2	KL3	KL4	KL5	KL6	KL7	KL8	KL9	KL10	KL11	KL12	KL13	KL14	KL15	KL16
1	.821	.429	-.370	-.241	-.275	-.410	-.197	-.119	.122	-.082	-.098	.043	.026	-.008	-.009	-.005
2	-.242	.232	-.417	.169	-.427	-.107	-.130	.007	-.235	.094	.120	-.056	-.043	-.005	.020	.006
3	.169	-.064	.471	.157	-.055	-.008	-.053	.178	-.028	-.115	-.097	-.003	-.059	-.055	-.000	-.017
4	-.457	-.068	-.316	.452	-.139	.172	.020	-.267	-.108	.031	-.081	.086	.007	.025	.003	-.008
5	.265	.559	-.261	-.237	.749	.004	-.024	-.013	-.179	.091	-.032	.016	-.035	-.032	-.004	.001
6	1.155	.274	-.098	.425	-.012	.262	-.069	.160	.298	.127	.050	-.037	-.017	-.000	.011	.001
7	-.317	-.044	.090	.183	.147	-.119	.076	-.018	.064	.020	.012	.064	.062	-.015	.023	-.011
8	-.174	-.167	.620	.339	.171	-.295	-.179	.032	-.082	.125	-.095	-.002	.014	.035	.009	.019
9	-.455	-.415	-.105	-.237	.115	.105	-.190	-.126	.079	.002	.077	-.070	.033	-.042	.007	.002
10	-.595	-.022	-.116	-.107	-.083	.146	.034	-.048	.118	.063	.002	.041	-.039	-.012	-.038	.021
11	-.278	-.026	-.107	.023	.158	.032	.126	-.101	.195	-.156	-.051	.077	-.058	.029	.023	.011
12	-.274	-.053	-.066	.143	.106	-.197	.139	-.081	.145	.045	-.031	-.141	.040	.016	-.037	-.003
13	.205	.041	-.067	-.160	-.256	.224	.106	.218	-.129	-.094	-.170	-.028	.063	-.018	.005	.021
14	1.325	-1.149	-.139	-.088	.053	.012	-.029	-.058	-.111	-.001	.007	.020	-.021	.003	-.010	-.001
15	-.619	-.109	-.022	-.197	.070	.133	-.179	.020	.059	.003	.000	-.004	.056	-.027	.027	-.006
16	-.282	-.250	-.132	-.342	-.114	-.274	.349	.188	.061	.129	.026	.003	-.029	.003	.028	-.003
17	.618	.364	.777	-.310	-.251	.122	.146	-.314	-.032	.061	.030	.017	.007	-.012	.002	.001
18	.475	.193	-.135	.053	.126	.110	.170	.115	-.137	-.023	.073	.020	.065	.063	-.011	-.008
19	-.290	.021	.150	.047	-.002	-.116	-.110	.210	.006	-.116	.222	.112	.019	-.004	-.025	.006
20	-.260	.034	-.004	.355	.010	-.044	.204	-.076	-.084	-.145	.033	-.067	-.017	-.065	-.011	-.004
21	-.098	.124	.134	-.192	.080	.046	-.102	-.042	-.020	-.195	.045	-.109	-.038	.087	.014	-.002
22	-.715	-.013	.106	-.243	-.169	.149	-.109	.136	-.003	.119	-.094	.019	-.034	.035	-.028	-.021

[a] KFn = Karhunen-Loève "n," where n signifies the number (*see* Table II) of the eigenvector from which that feature is derived.

Table VI. Varimax Features[a]

STATION	VM1	VM2	VM3	VM4	VM5	VM6	VM7	VM8	VM9	VM10	VM11	VM12	VM13	VM14	VM15	VM16
1	-.287	.724	.084	.684	-.563	-.851	.699	.342	-.430	.745	-.316	-.037	-.274	-.131	.026	.021
2	.298	-.352	-.376	.333	-.083	-.057	.106	-.087	.001	-.095	.030	-.130	.068	.064	-.006	-.060
3	-.047	.165	.512	.059	-.130	.071	-.174	.155	-.034	.102	.325	.090	-.020	-.019	.022	.057
4	.257	-.546	-.390	.107	.392	.285	-.189	-.038	.377	-.267	-.310	-.122	.064	.034	.001	-.073
5	.154	.628	-.153	.026	.160	-.199	.827	.024	-.172	.137	.046	-.140	.003	-.207	-.009	-.039
6	-.561	1.010	.546	.946	-.356	-.652	.764	1.127	-.468	1.079	-.537	.177	-.214	-.109	.038	.112
7	.219	-.280	-.016	-.219	.334	.143	-.200	-.151	.269	-.213	.046	.054	.063	.061	.011	-.068
8	.109	-.199	.605	-.243	.317	.242	-.304	-.110	.175	-.208	.032	-.001	.086	.014	-.013	.043
9	-.042	-.485	-.344	-.483	.270	.393	-.246	-.478	.269	-.292	.190	.030	-.073	-.012	-.097	-.026
10	.357	-.540	-.452	-.389	.140	.380	-.302	-.336	.312	-.329	.206	-.036	.055	-.007	-.032	-.019
11	.164	-.201	-.254	-.214	.285	.127	-.157	-.158	.210	-.094	-.012	.071	.066	-.020	.036	.019
12	.168	-.270	-.128	-.167	.328	-.013	-.153	-.140	.205	-.135	.007	-.116	.032	.018	-.032	.025
13	-.099	.210	-.043	.214	-.438	-.085	.046	.119	-.075	.119	-.018	-.071	-.127	.046	.055	.011
14	-1.750	.827	.634	.596	-.279	-.793	.371	.542	-.406	.607	-.796	.231	-.163	-.215	.103	.168
15	.303	-.533	-.381	-.512	.193	.538	-.268	-.494	.352	-.335	.322	.021	-.057	.048	-.080	-.079
16	-.010	-.312	-.282	-.358	-.067	-.183	-.268	-.389	.191	-.322	.248	-.025	.160	-.008	.159	-.026
17	-.112	.731	.717	.064	-.700	-.063	-.059	.612	-.837	.116	.044	-.182	-.017	.013	.049	-.003
18	-.194	.522	.102	.385	-.155	-.328	.399	.376	-.143	.245	-.141	.032	.038	.083	.005	-.014
19	.236	-.256	.028	-.171	.105	.184	-.184	-.239	.179	-.147	.330	.297	.111	.170	-.101	-.038
20	.240	-.242	-.058	.054	.324	.095	-.233	.003	.267	-.218	-.044	-.043	.106	.071	.001	-.010
21	.144	.028	-.017	-.170	-.020	.199	-.059	-.159	-.087	-.058	.242	-.021	.041	.081	-.087	.045
22	.453	-.631	-.336	-.546	-.029	.566	-.412	-.520	.349	-.441	.407	-.083	.059	.023	-.048	-.044

[a] VFn = Varimax feature "n," where n signifies the varivector (*see* Table III) from which that feature is derived.

map; and second, the pleasing appearance is aided by the fact that the mapping program fits contour lines in such a manner that contour lines which are diverging at the edge of the data space continue to diverge to the edge of the map space. Of major importance is the fact that the second eigenvector is constrained to be orthogonal to the first, whereas the varivectors have no such constraint. Hence, the map of the first varimax feature is surely a better representation of the smelter "plume" than the map of the second Karhunen-Loève feature. Further lending credence to this assumption is a comparison of these maps with the maps of arsenic, antimony, and copper, Figures 2, 3, and 4.

A similar situation can be noted in comparison of the maps for the third Karhunen-Loève and varimax features, both of which consist primarily of nitric acid. The map of the third Karhunen-Loève feature is somewhat "spotty" (due, perhaps in part, to the orthogonality constraint), whereas the map of the third varimax feature, while similar, shows two large areas of increased concentration of both hydrogen and nitrate ions, and is not "spotty." These areas of increased nitric acid concentration occur downwind of heavily urbanized and industrialized areas of the Puget Sound region.

Maps of the remaining features are not included, since, as has been noted already, the vectors from which these features are derived consist mainly of single elements, and are not readily identifiable as real factors in the data set.

Summary and Conclusions. In summary, it has been shown that by analyzing for a variety of species in rain samples gathered over a sufficiently large geographical area, a grasp of some of the influences on atmospheric chemistry, as reflected in precipitation, is obtainable. Analysis of the data appears to show at least three major sources of dissolved species in rainwater in the Puget Sound region: a sea-salt background, urban sources, and an industrial source of major significance. Mediating these sources are such variables as time/speed/direction wind patterns and time/volume precipitation patterns, which in turn are probably influenced by geographic features and by the aforementioned sources themselves (e.g., inputs of heat, particulates, and other chemical entities into the atmosphere).

Varimax rotation of the eignevector matrix derived from the correlation matrix has been demonstrated to more accurately represent the major factors in the chemical composition of the rain samples studied. Further, these studies have revealed the varimax-rotated vector representation to be considerably more stable to the effects of analytical errors, than the eignevector representation.

This appears to have been the first application of both performing a rotation analogous to the Karhunen-Loève transform by using the varivector matrix, and of mapping the resulting features over a geographical data space. The application of these techniques to similar problems may reasonably be expected to be a useful tool in understanding some of the complex interactions encountered in similar (e.g., environmental) research in the future.

While the number of data gathered in this study were limited, they were nonetheless sufficient to demonstrate the utility of the approach for studying sources and influences on atmospheric chemical composition.

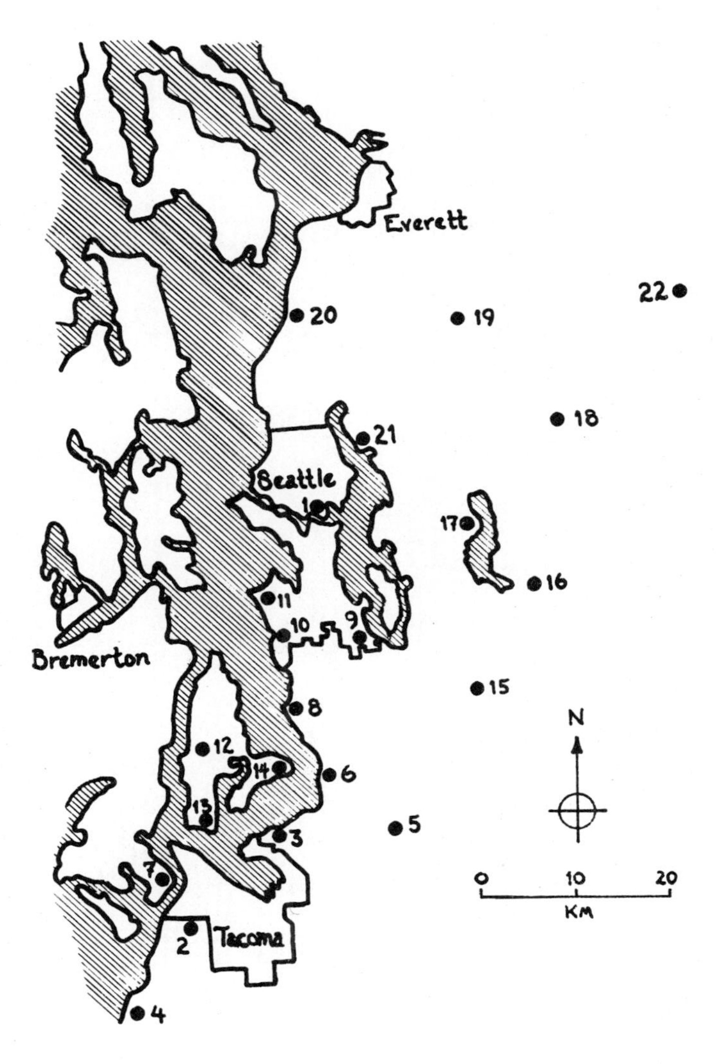

Figure 1. Sampling sites. ● = *Samping site.*

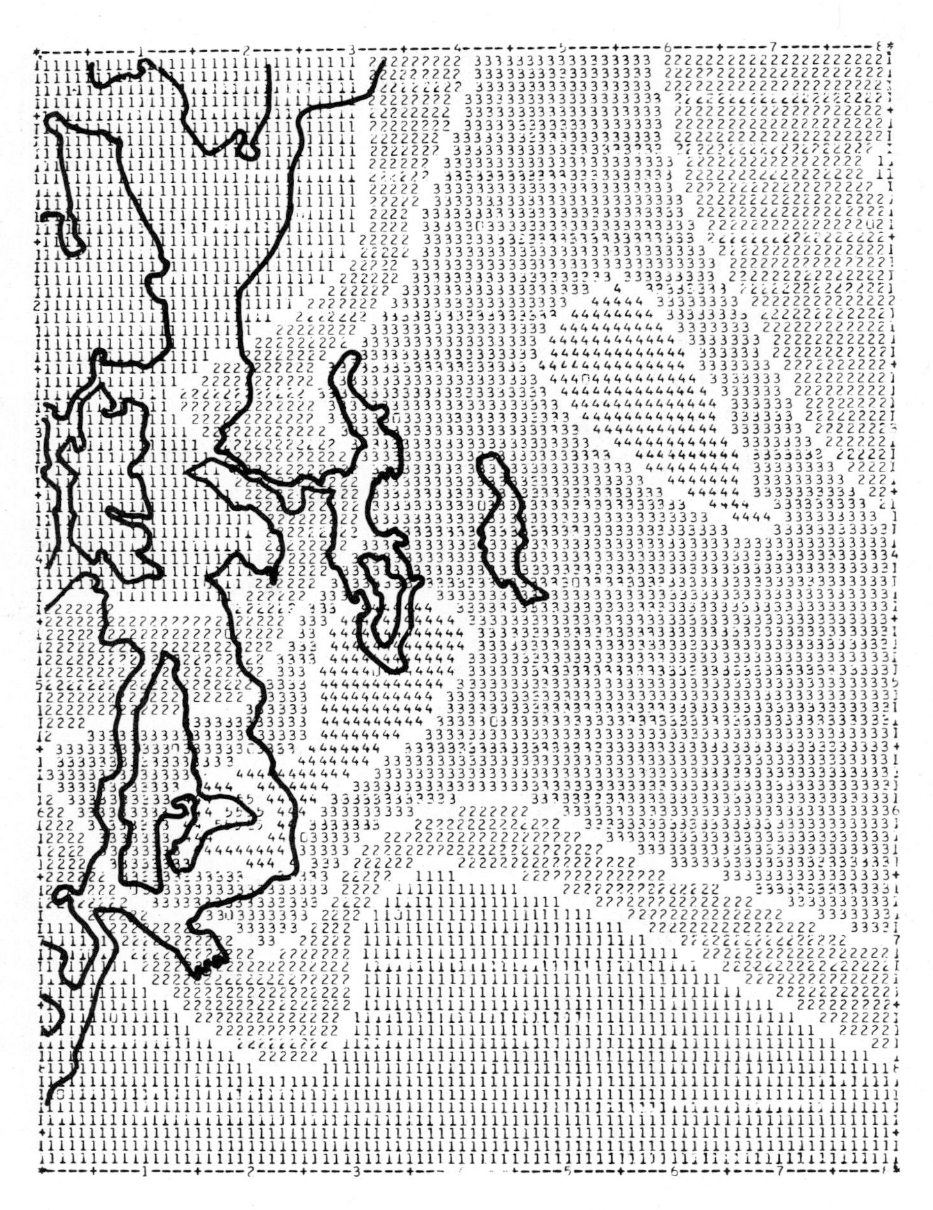

Figure 2. Map of arsenic concentrations. Contour levels: 1 = 0.6–1.4 ppb; 2 = 1.4–3.4 ppb; 3 = 3.4–8.2 ppb; 4 = 8.2–19.8 ppb; 5 = 19.8–47 ppb. 0 = sampling site.

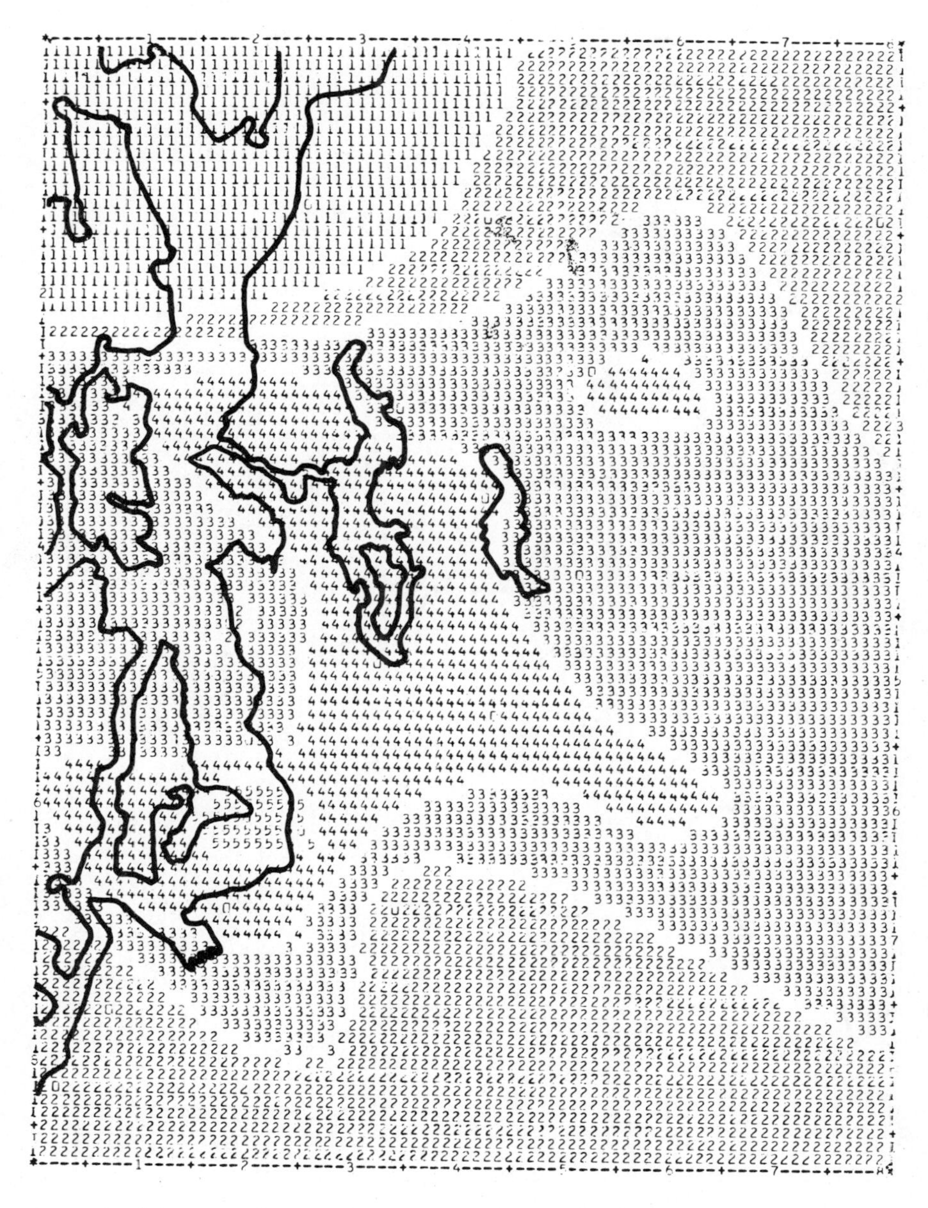

Figure 3. Map of antimony concentrations. Contour levels: 1 = 0.0–0.07 ppb; 2 = 0.07–0.16 ppb; 3 = 0.16–0.38 ppb; 4 = 0.38–0.85 ppb; 5 = 0.85–2.1 ppb. 0 = sampling site.

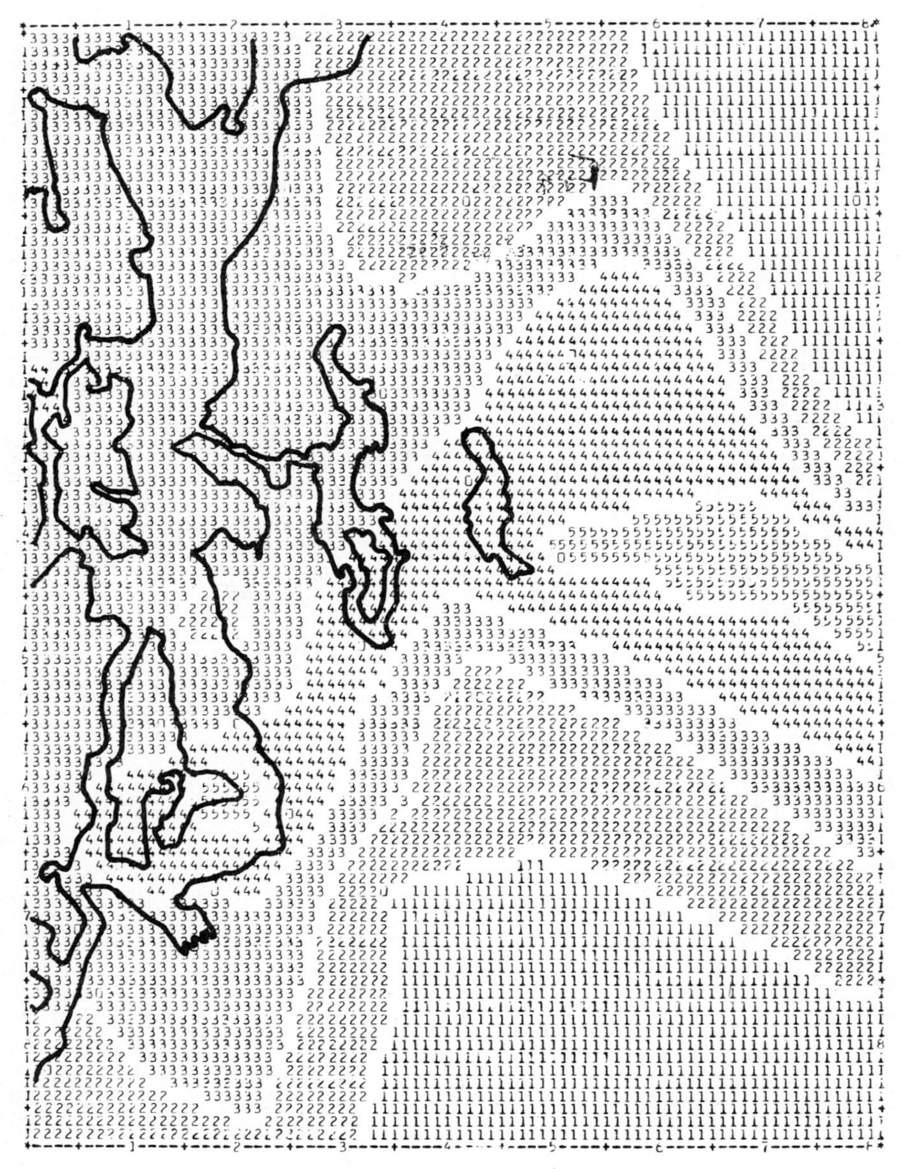

Figure 4. Map of copper concentrations. Contour levels: 1 = 2–5.4 ppb; 2 = 5.4–15 ppb; 3 = 15–40 ppb; 4 = 40–108 ppb; 5 = 108–290 ppb. 0 = sampling site.

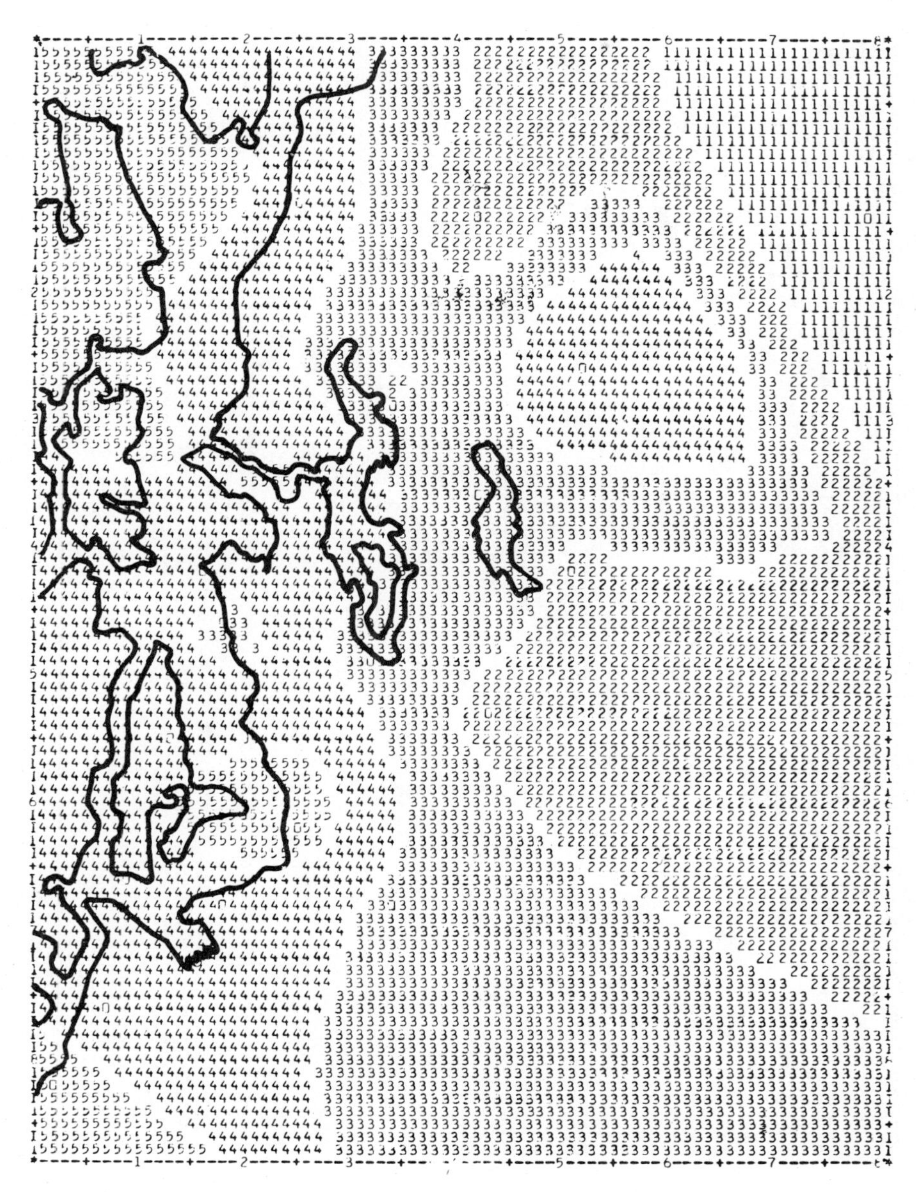

Figure 5. Map of cadmium concentrations. Contour levels: 1 = 0.0–0.04 ppb; 2 = 0.04–0.09 ppb; 3 = 0.09–0.20 ppb; 4 = 0.20–0.43 ppb; 5 = 0.43–0.92 ppb. 0 = sampling site.

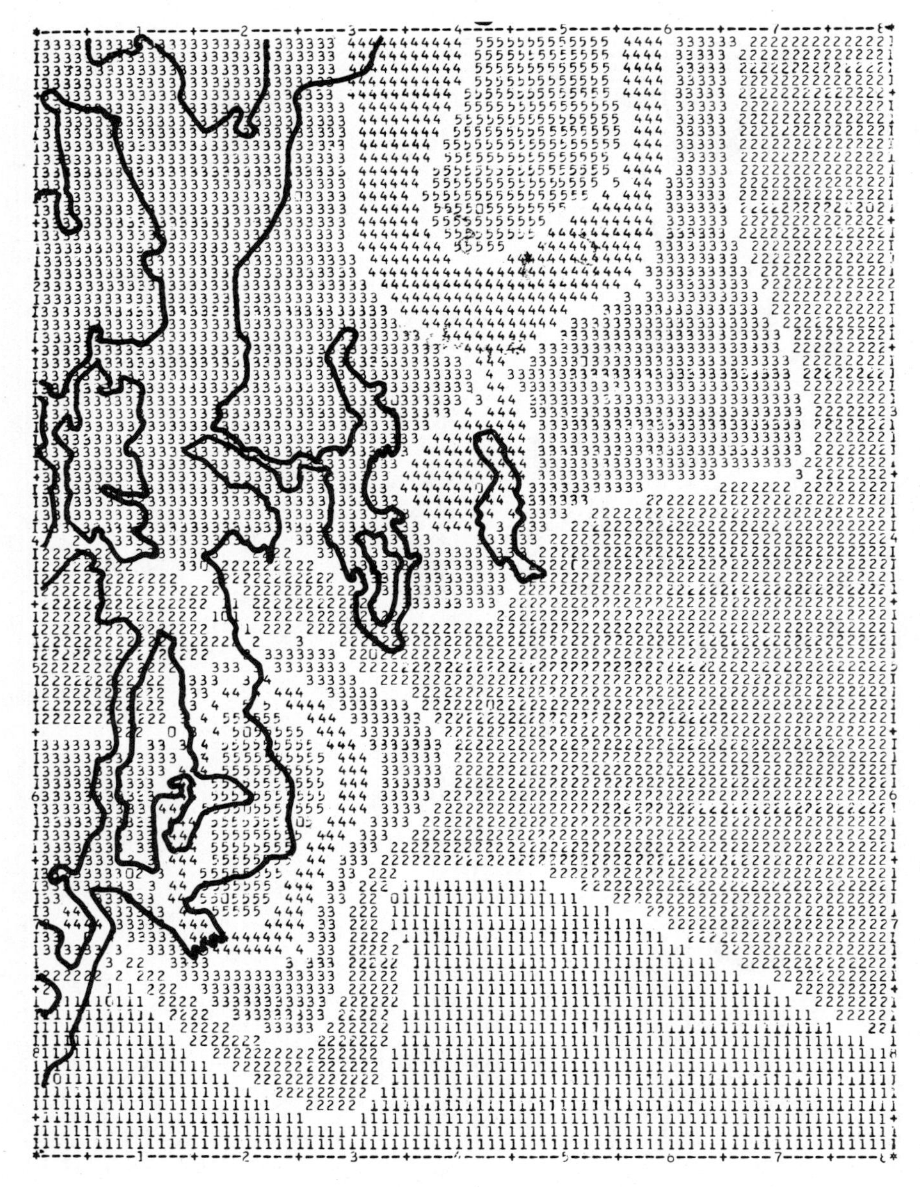

Figure 6. Map of pH. Contour levels: 1 = 4.59–4.75; 2 = 4.43–4.59; 3 = 4.27–4.43; 4 = 4.11–4.27; 5 = 3.95–4.11. 0 = sampling site.

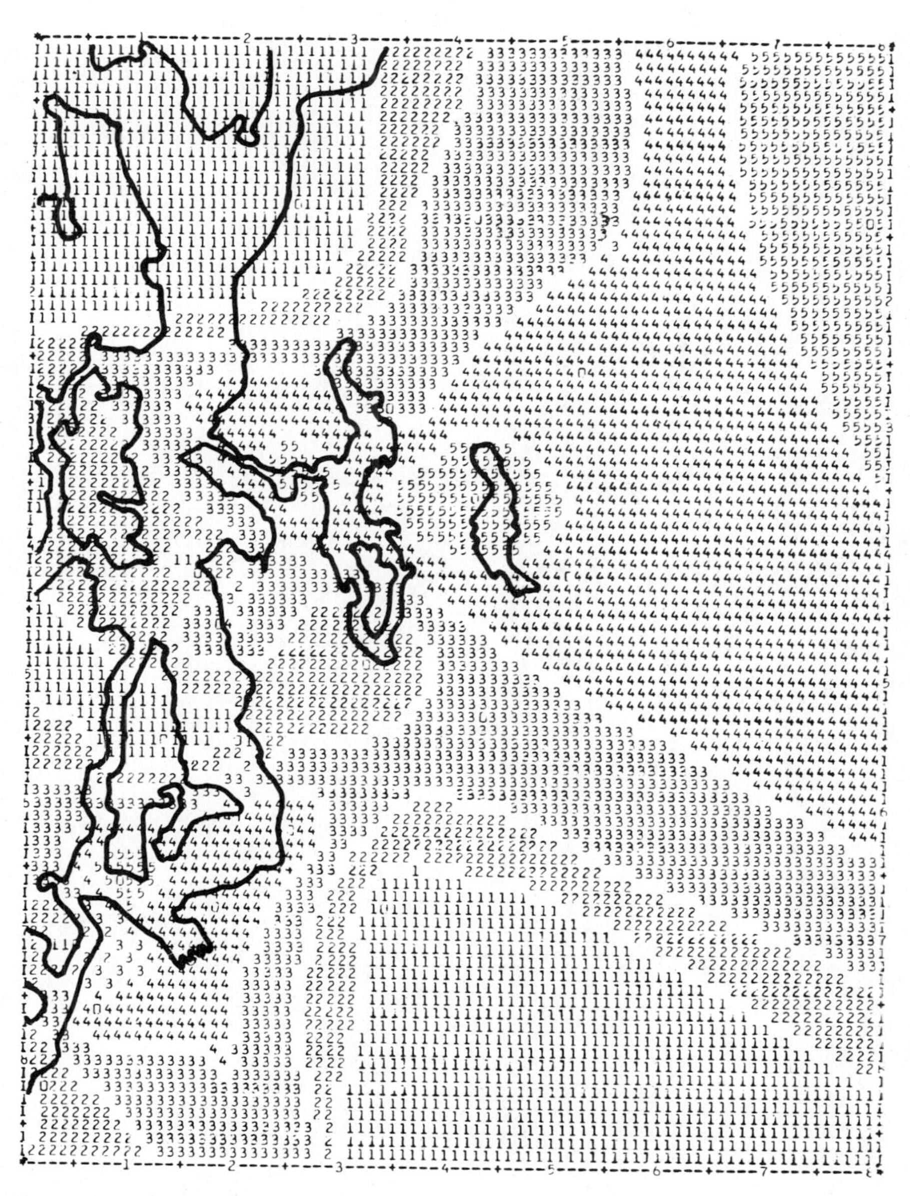

Figure 7. Map of sulfate concentrations. Contour levels: 1 = 1100–1450 ppb; 2 = 1450–1900 ppb; 3 = 1900–2500 ppb; 4 = 2500–3200 ppb; 5 = 3200–4300 ppb. 0 = sampling site.

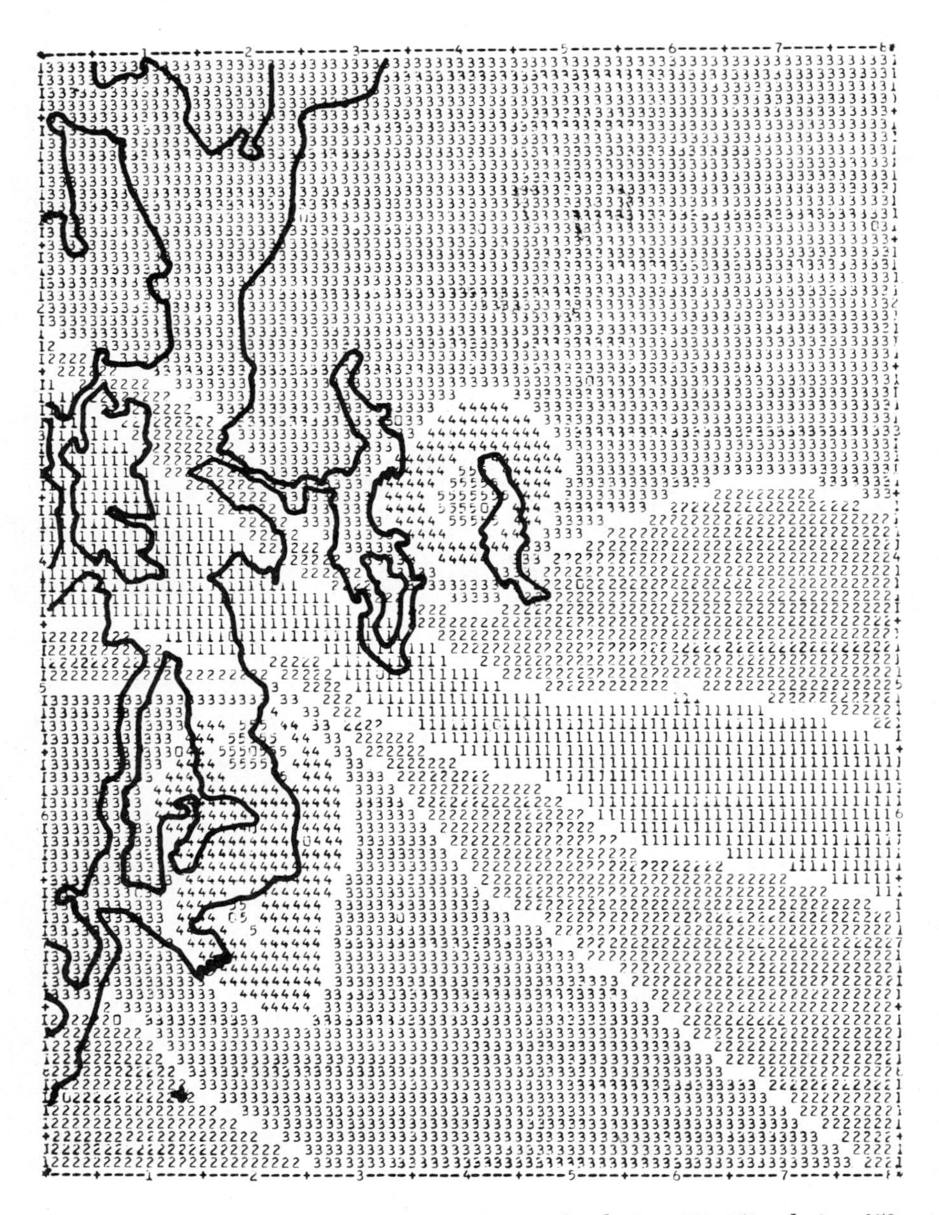

Figure 8. Map of nitrate concentrations. Contour levels: 1 = 170–250 ppb; 2 = 250–380 ppb; 3 = 380–560 ppb; 4 = 560–840 ppb; 5 = 840–1250 ppb. 0 = sampling site.

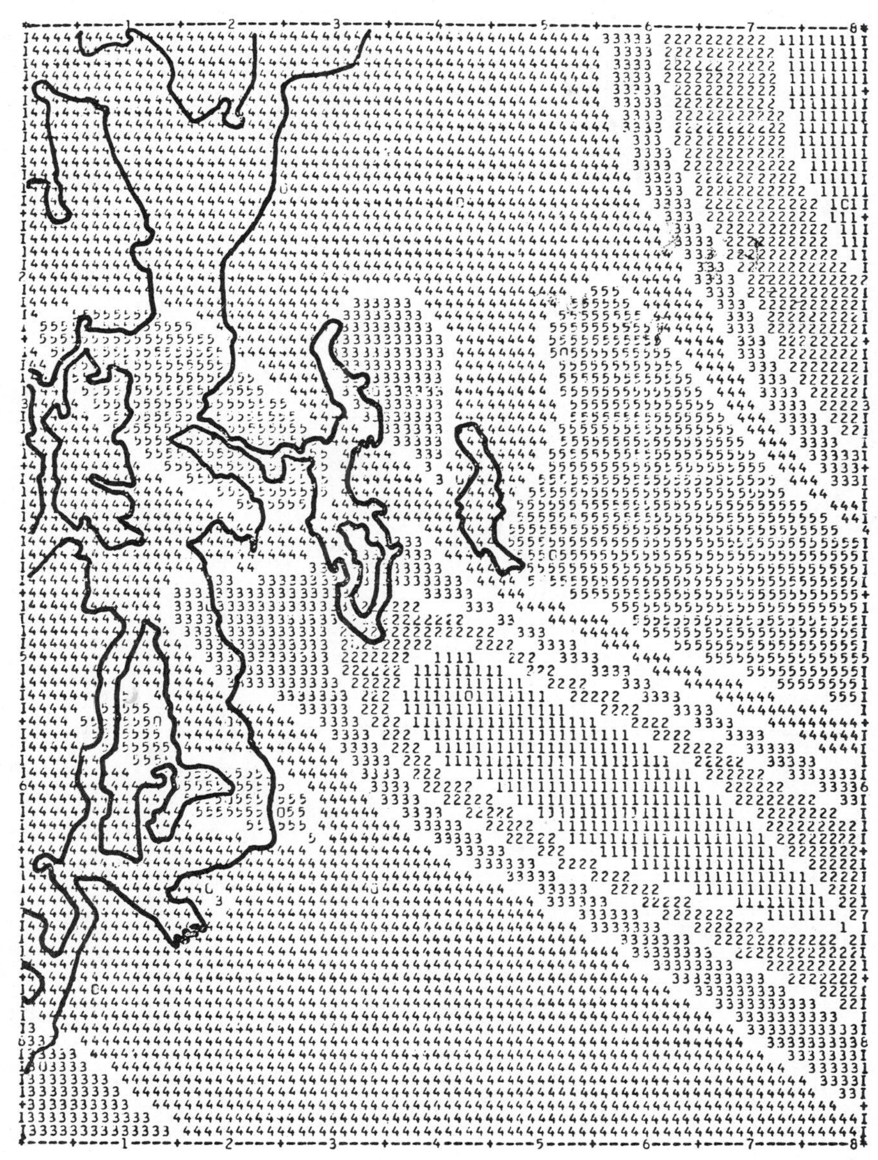

Figure 9. Map of zinc concentrations. Contour levels: 1 = 1.5–2.7 ppb; 2 = 2.7–4.9 ppb; 3 = 4.9–8.9 ppb; 4 = 8.9–16 ppb; 5 = 16–29 ppb. 0 = sampling site.

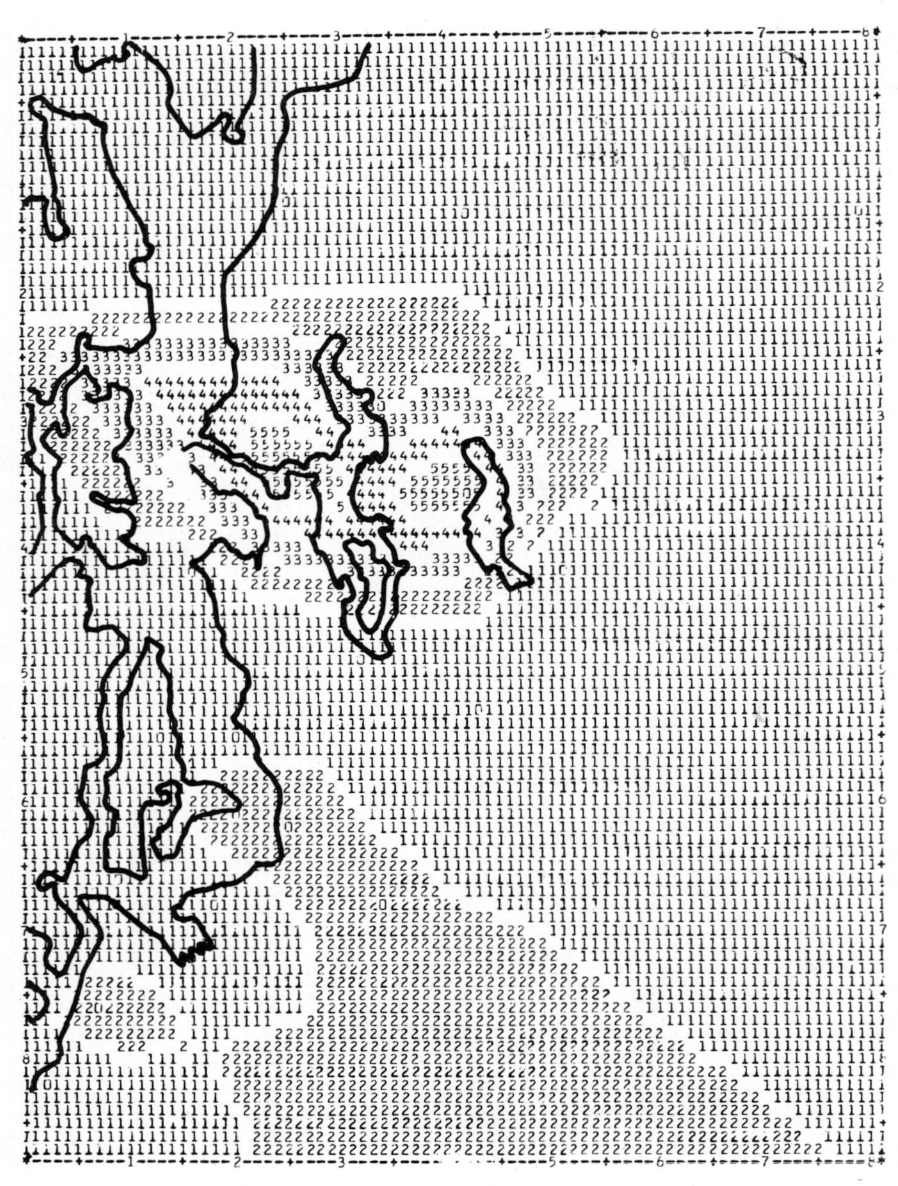

Figure 10. Map of lead concentrations. Contour levels: 1 = 3–14 ppb; 2 = 14–25 ppb; 3 = 25–36 ppb; 4 = 36–47 ppb; 5 = 47–58 ppb. 0 = sampling site.

	H+	NH4+	NA	K	CA	MG	ZN	CU	PB	SO4	MN	CD	AS	SB	CL-	NO3-
H+	1	-.10	.49	.37	.31	.36	.13	.40	.15	.05	.10	.29	.41	.40	.27	.74
NH4+		1	.46	.24	.61	.48	.41	.05	.26	.03	.42	.24	.07	.23	.68	-.11
NA			1	.81	.71	.94	.47	.47	.52	.34	.33	.39	.39	.57	.92	.47
K			.81	1	.64	.61	.24	.19	.33	.29	.50	.48	.09	.28	.49	.43
CA			.71		1	.61	.56	.21	.48	.33	.62	.48	.14	.52	.66	.18
MG			.94			1	.39	.27	.54	.38	.33	.22	.19	.36	.88	.42
ZN							1	.46	.38	.21	.49	.44	.26	.41	.45	.05
CU								1	.14	.23	.11	.62	.94	.85	.36	.14
PB									1	.52	.27	.11	.10	.33	.53	.35
SO4										1	.29	-.04	.12	.35	.26	.16
MN											1	.55	.03	.24	.50	.09
CD												1	.55	.64	.32	.12
AS								.94					1	.85	.30	.09
SB								.85					.85	1	.50	.12
CL-			.92			.88									1	.31
NO3-	.74															1

Figure 11. Correlation matrix. Values in lower left half are correlations ≥ 0.7.

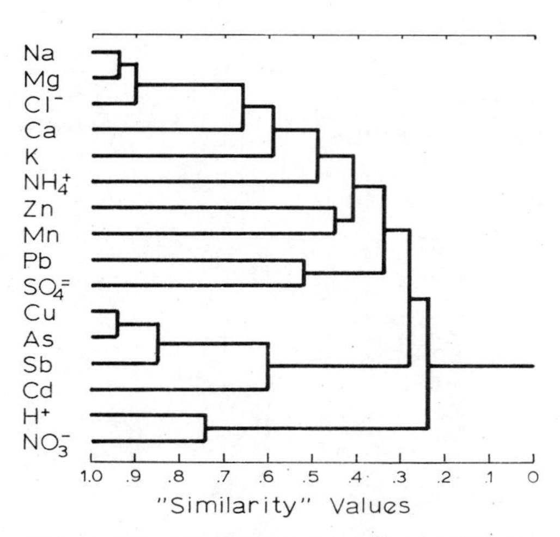

Figure 12. Hierarchical dendogram. "Similarity values" are absolute values of correlations or averaged correlations.

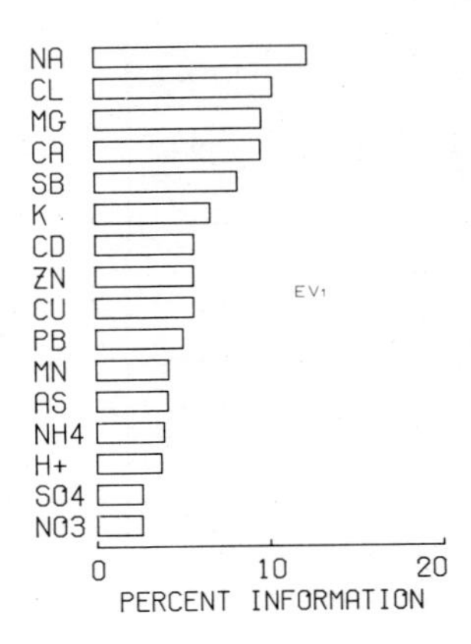

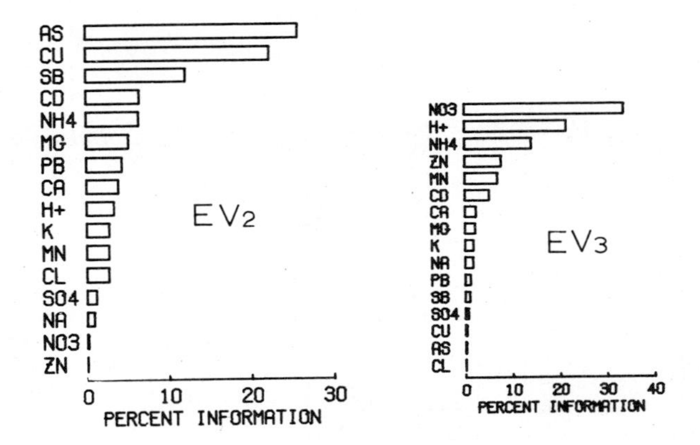

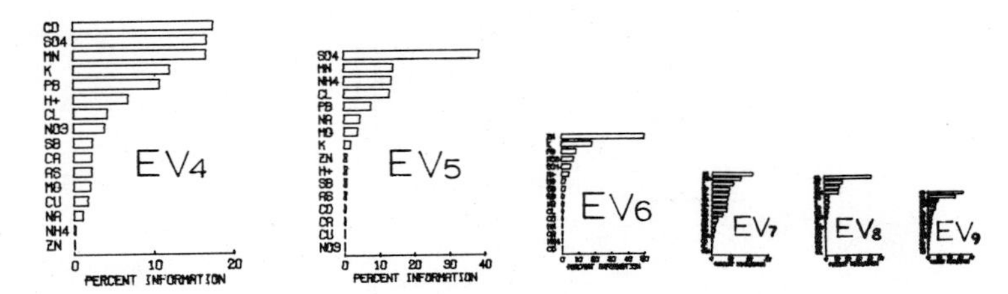

Figure 13. Histograms of eigenvectors. EV_n = *Eigenvector "n," where "n" is the magnitude-rank of the eigenvalue* (see *Table II*). *"Percent information" is equivalent to the loading for a particular species divided by the sum of the loadings in that eigenvector.*

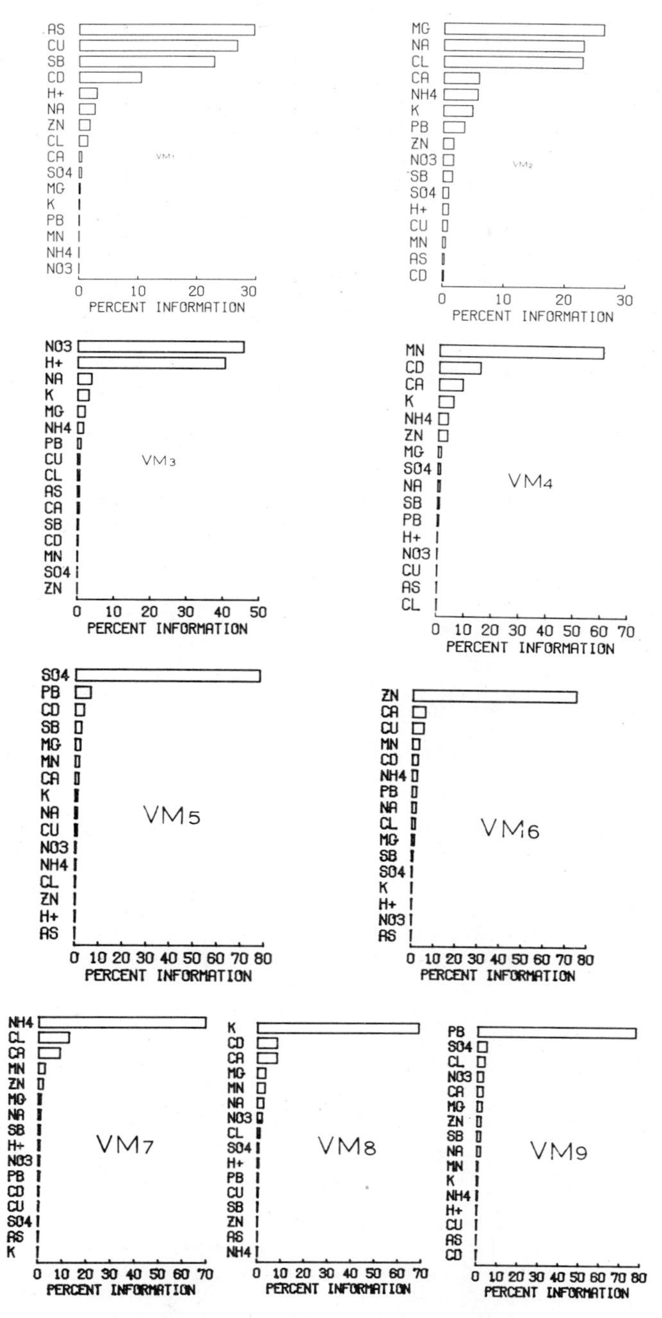

Figure 14. Histograms of varivectors. VM_n = Varimax vector "n." Interpretation similar to Figure 13.

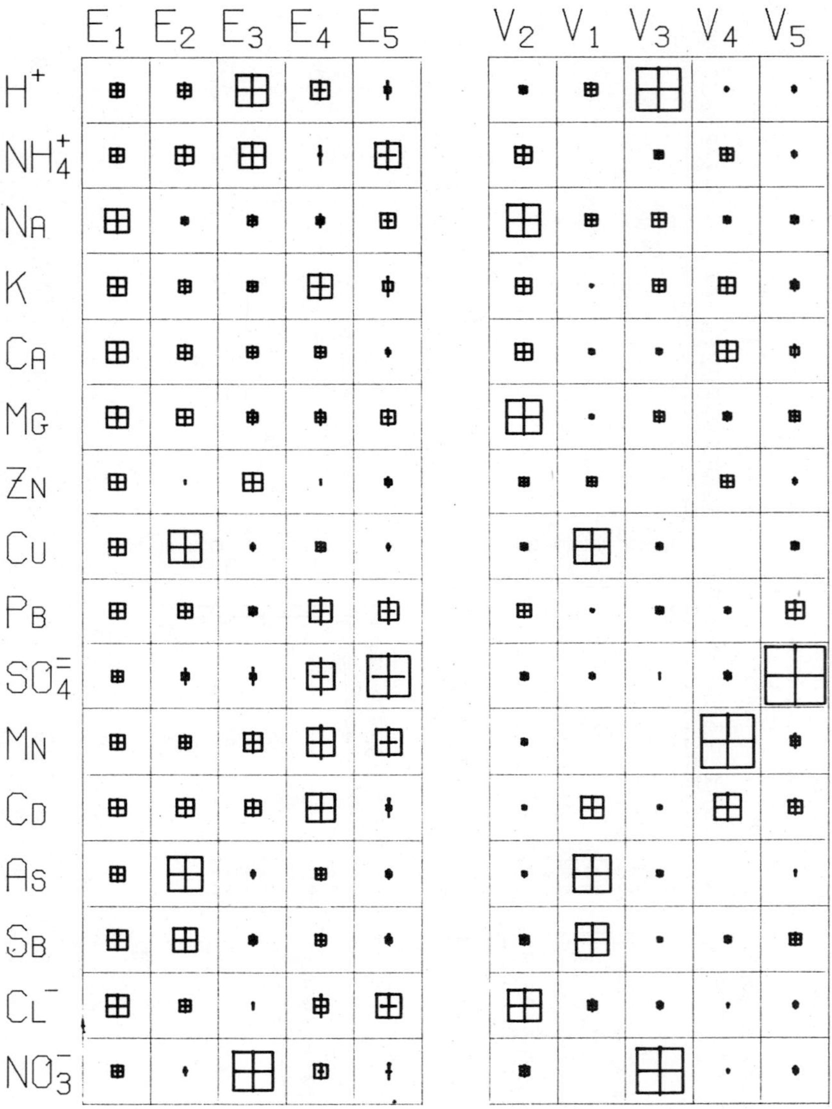

E_1 E_2 E_3 E_4 E_5 V_2 V_1 V_3 V_4 V_5

H^+, NH_4^+, Na, K, Ca, Mg, Zn, Cu, Pb, $SO_4^=$, Mn, Cd, As, Sb, Cl^-, NO_3^-

Figure 15. Comparison of eigenvectors and varivectors, including error data. E_n = Eigenvector "n," where n *is the magnitude-rank of the eigenvector* (see *Table II*). V_n = *Varivector "n," where* n *is the magnitude-rank of the varivector* (see *Table III*). *Note: Varivectors are not ordered by varivalue ranking in this figure—rather, they are ordered by the closest correspondence to the eigenvectors in terms of the "identity" of the vectors, based on their relative loadings.*

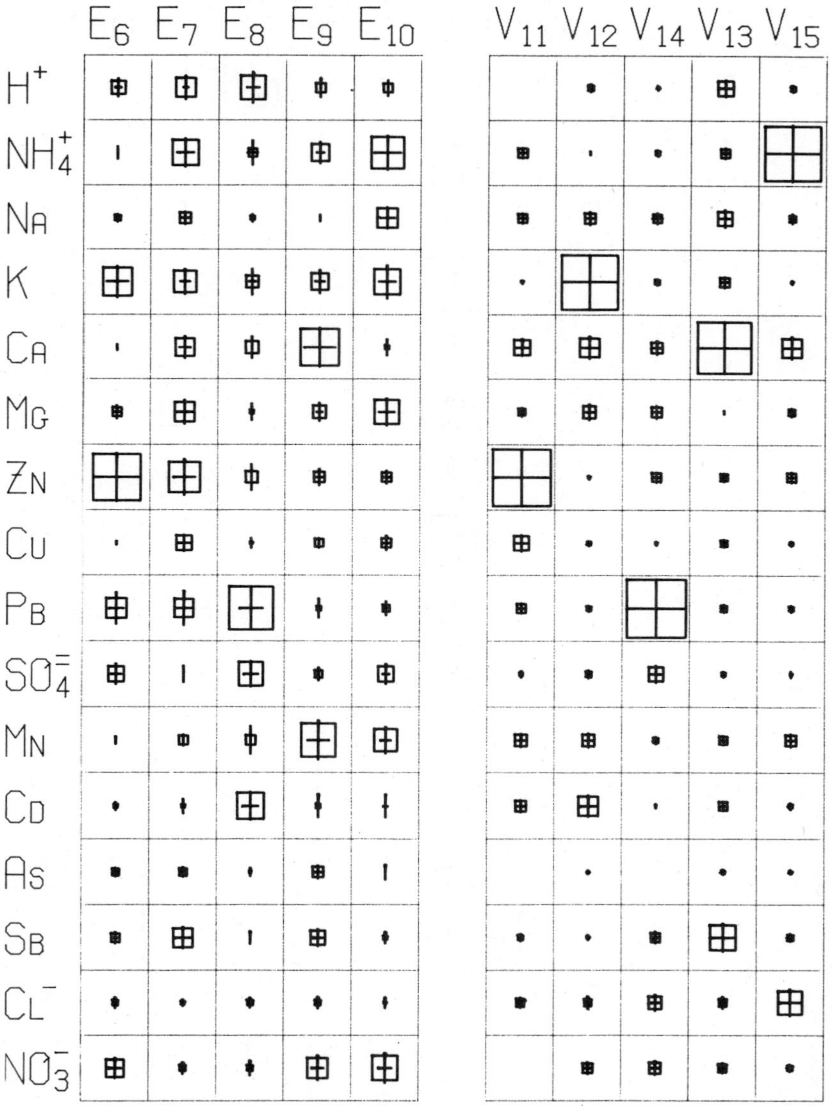

E_6 E_7 E_8 E_9 E_{10} V_{11} V_{12} V_{14} V_{13} V_{15}

H^+
NH_4^+
Na
K
Ca
Mg
Zn
Cu
Pb
$SO_4^=$
Mn
Cd
As
Sb
Cl^-
NO_3^-

Figure 15 (*continued*)

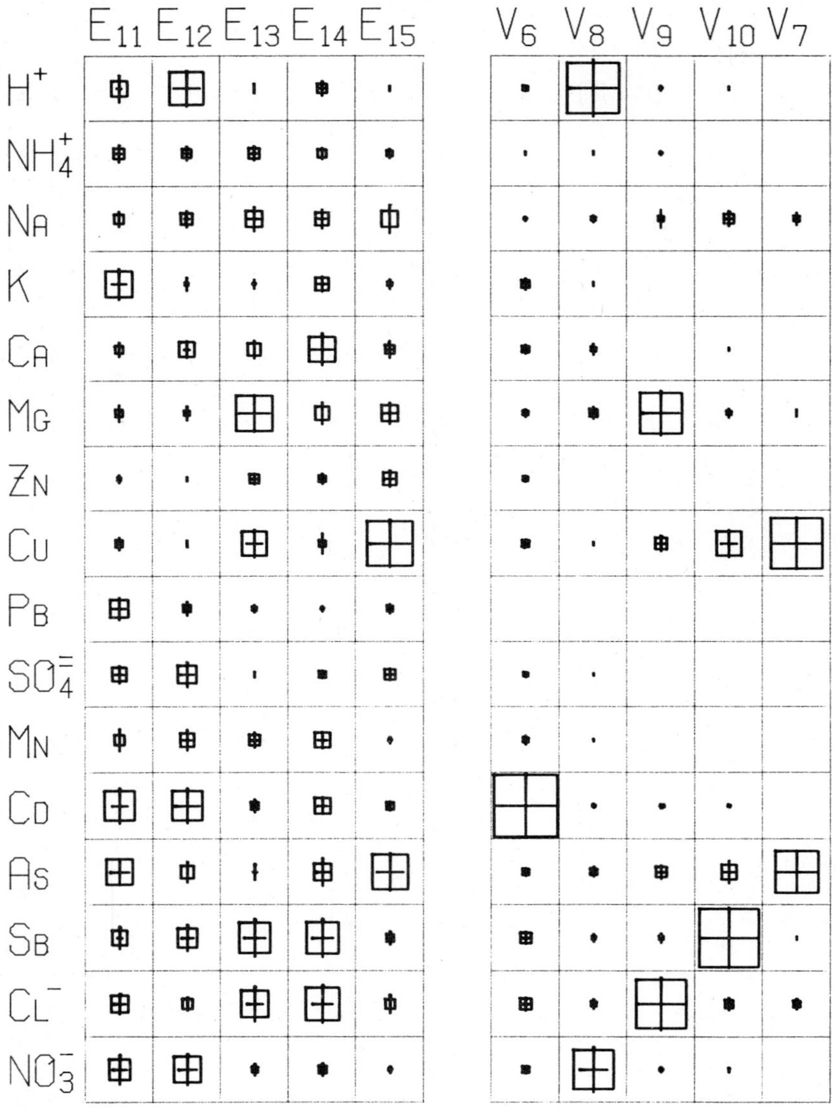

E_{11} E_{12} E_{13} E_{14} E_{15} V_6 V_8 V_9 V_{10} V_7

H^+
NH_4^+
Na
K
Ca
Mg
Zn
Cu
Pb
$SO_4^=$
Mn
Cd
As
Sb
Cl^-
NO_3^-

Figure 15 (*continued*)

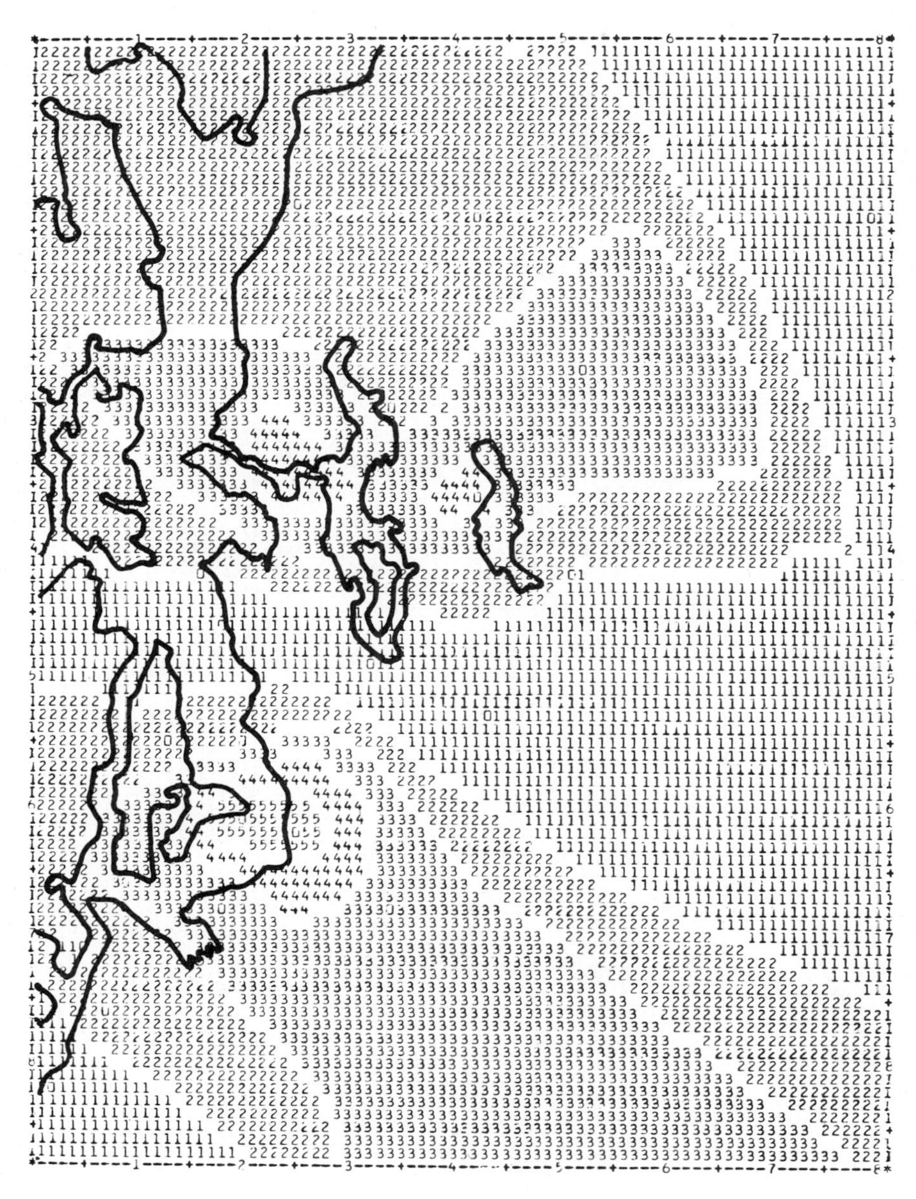

Figure 16. Map of first Karhunen–Loève feature. Contour levels: 1 = −0.715–−0.306; 2 = −0.306–0.103; 3 = 0.103–0.512; 4 = 0.512–0.921; 5 = 0.921–1.330. 0 = sampling site.

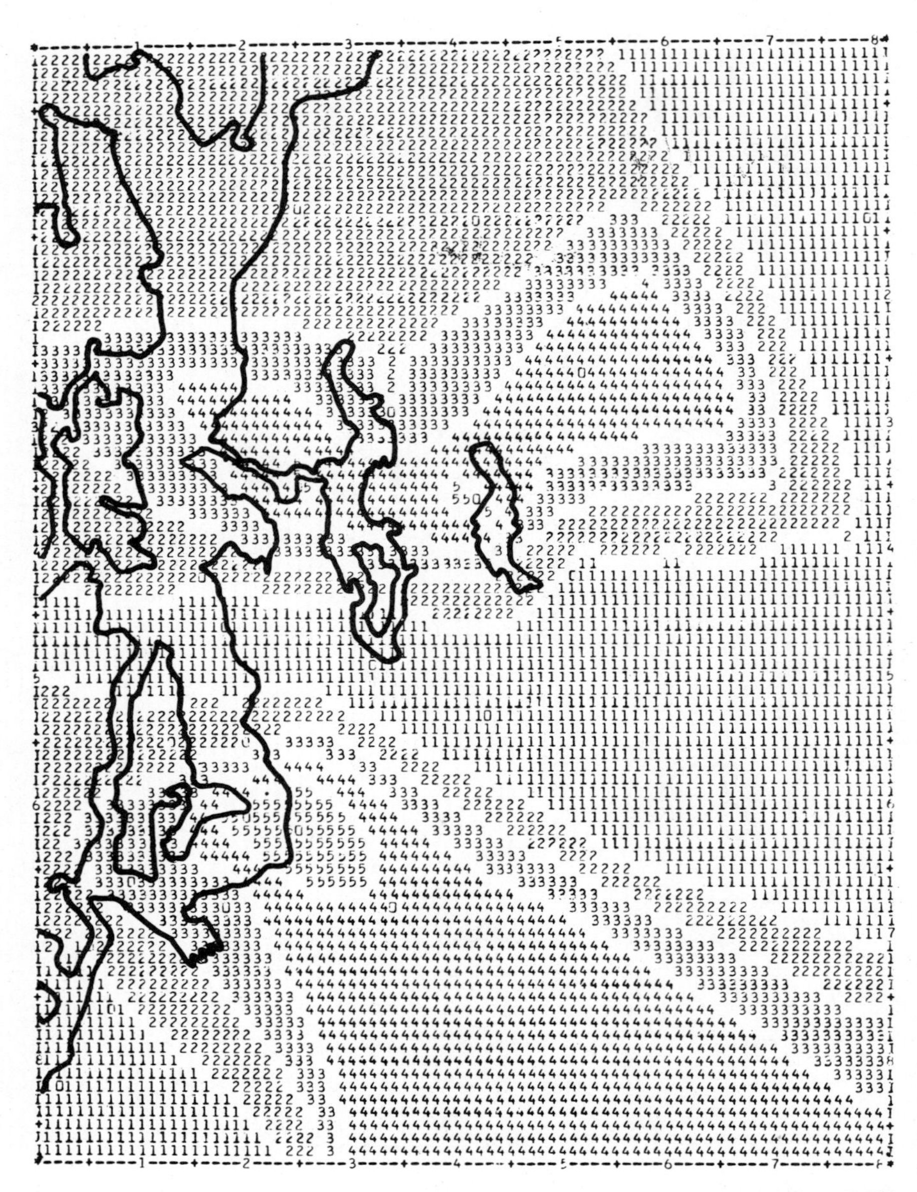

Figure 17. Map of second varimax feature. Contour levels: 1 = −0.631– −0.303; 2 = −0.303–0.025; 3 = 0.025–0.353; 4 = 0.353–0.682; 5 = 0.682–1.010. 0 = sampling site.

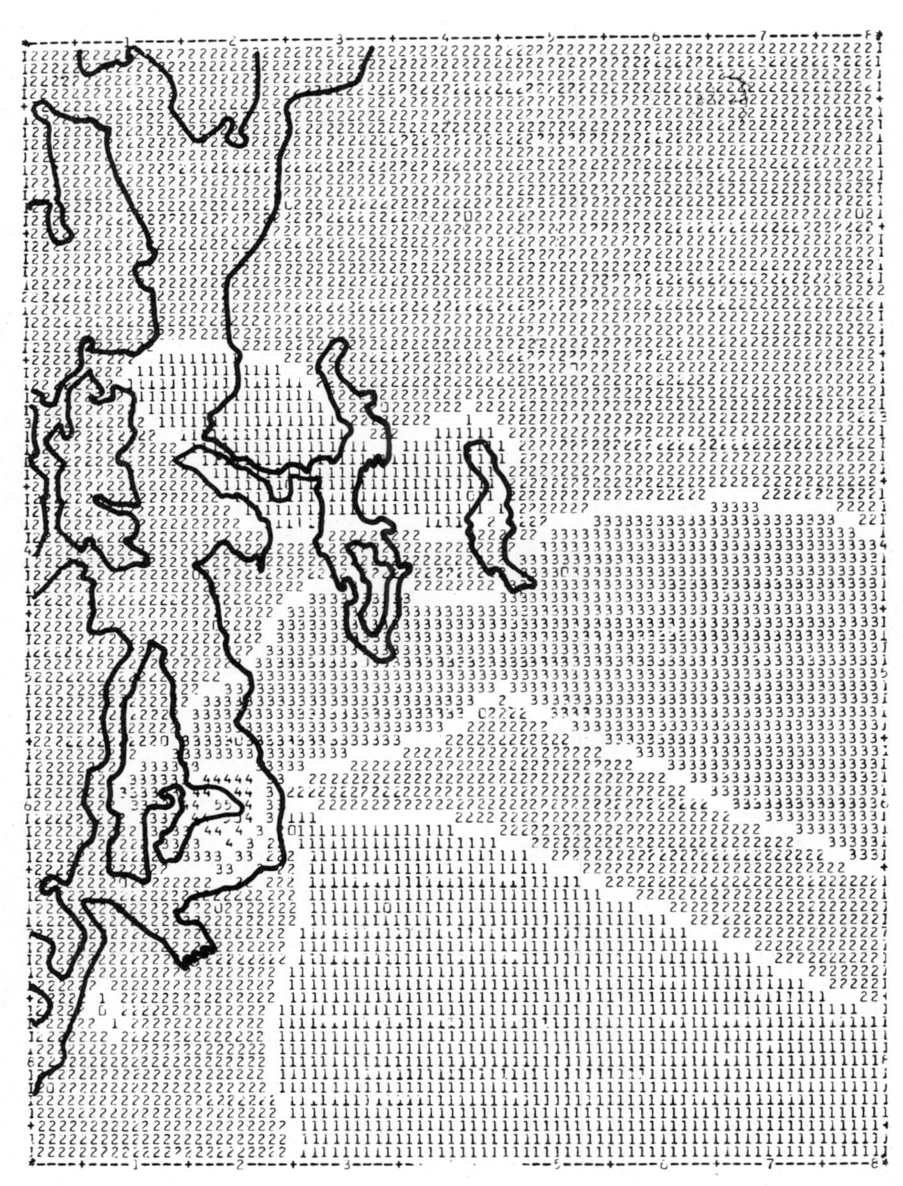

Figure 18. Map of second Karhunen–Loève feature. Contour levels: 1 = 0.559–0.217; 2 = 0.217– –0.125; 3 = –0.125– –0.466; 4 = –0.466– –0.808; 5 = –0.808– –1.150. 0 = sampling site.

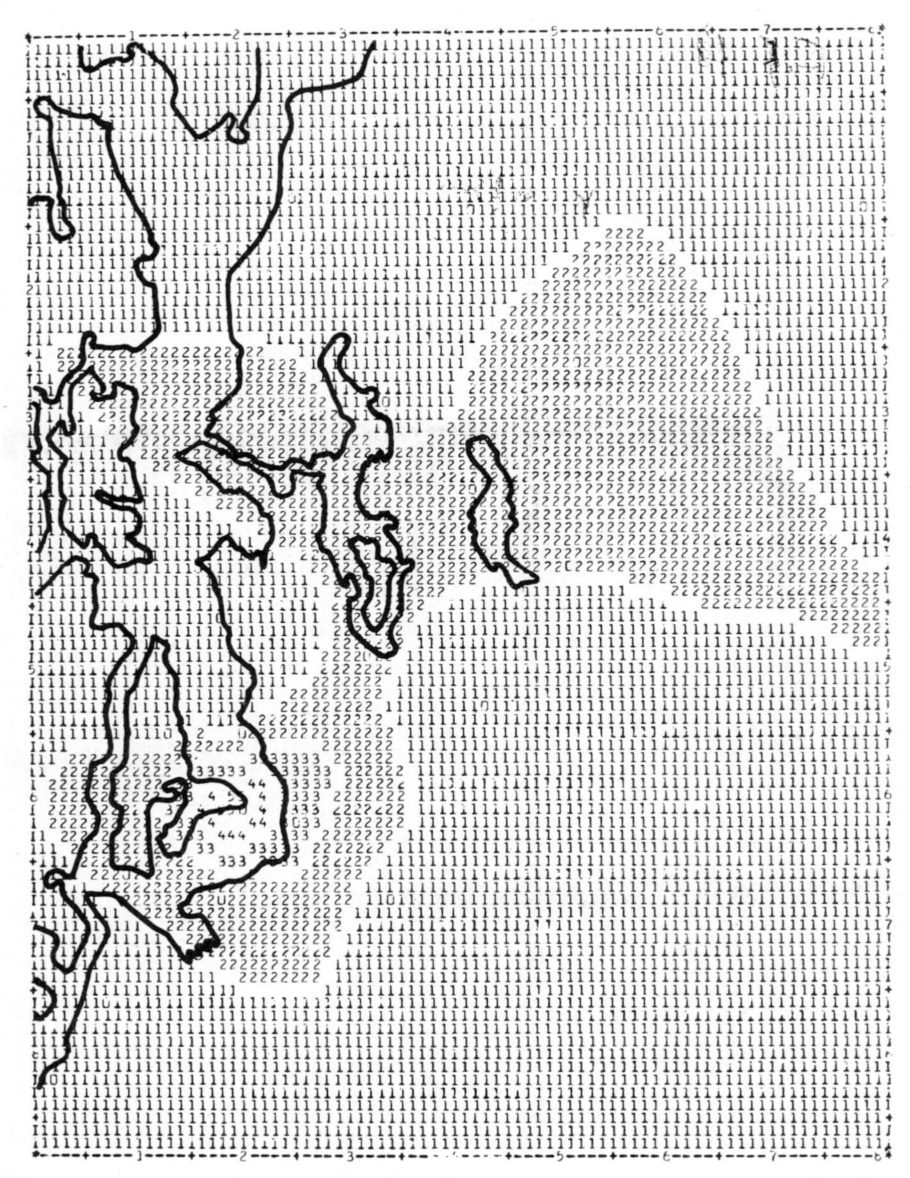

Figure 19. Map of first varimax feature. Contour levels: 1 = 0.453–0.012; 2 = 0.012– −0.428; 3 = −0.428– −0.869; 4 = −0.869– −1.309; 5 = −1.309– −1.750. 0 = sampling site.

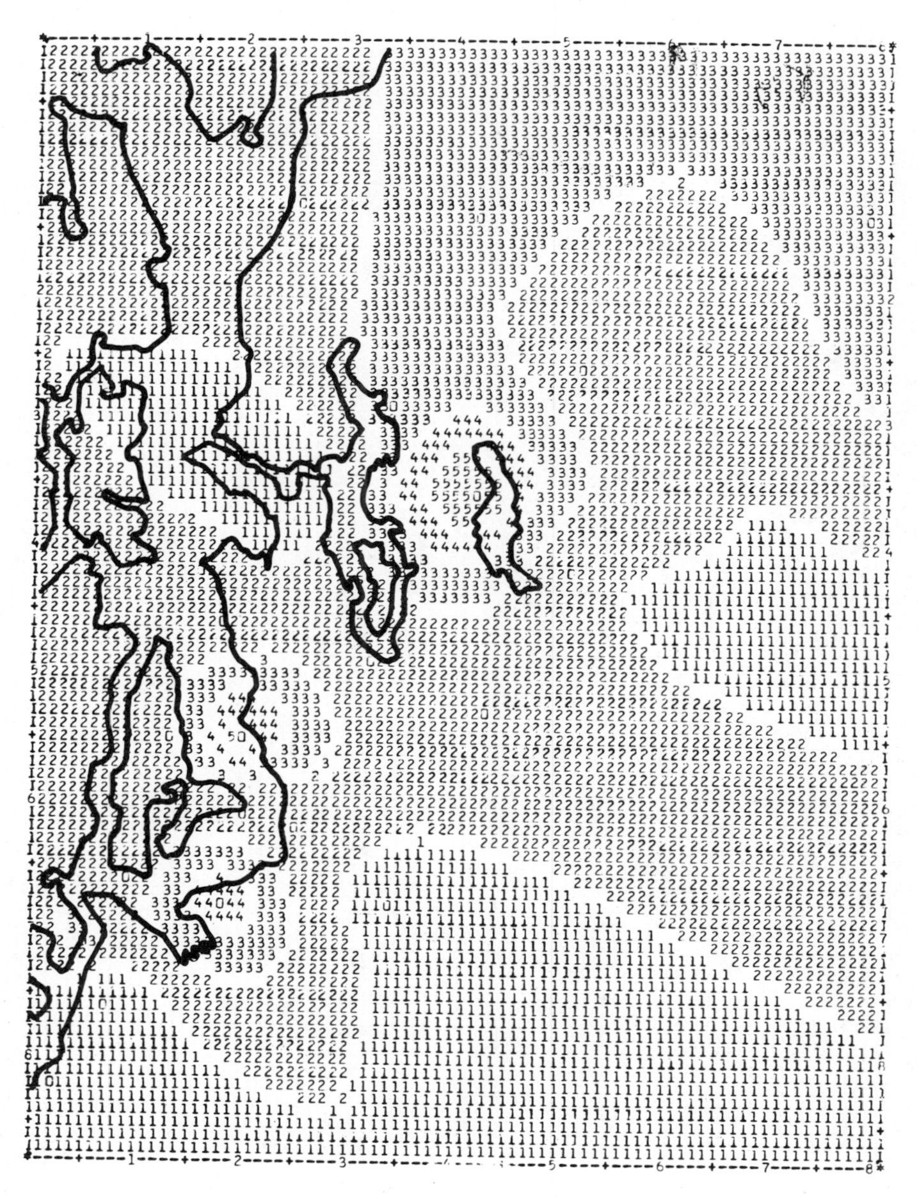

Figure 20. Map of third Karhunen–Loève feature. Contour levels: 1 = −0.417–−0.178; 2 = −0.178–0.061; 3 = 0.061–0.229; 4 = 0.299–0.538; 5 = 0.538–0.777. 0 = sampling site.

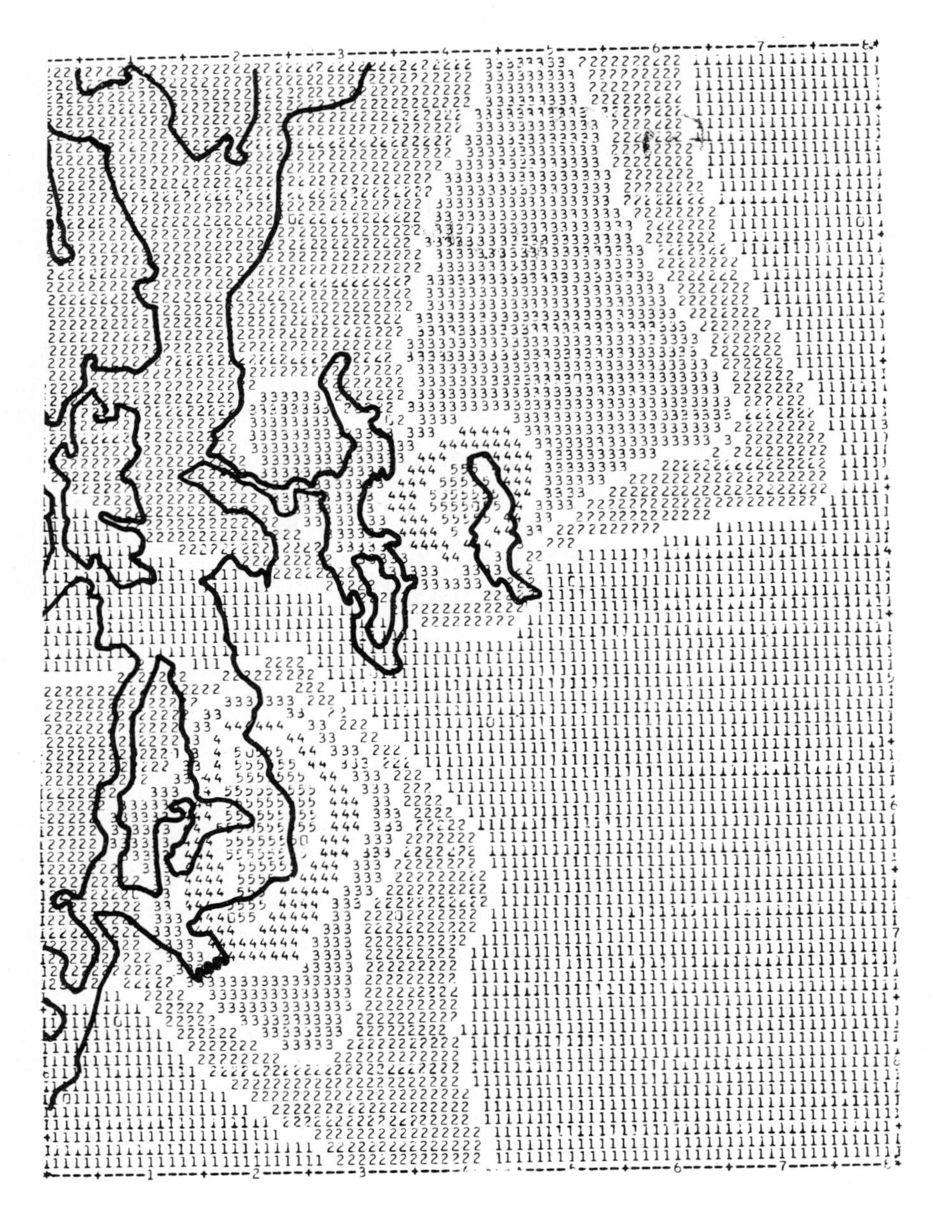

Figure 21. Map of third varimax feature. Contour levels: 1 = −0.452– −0.218; 2 = −0.218–0.016; 3 = 0.016–0.249; 4 = 0.249–0.483; 5 = 0.483–0.717. 0 = sampling site.

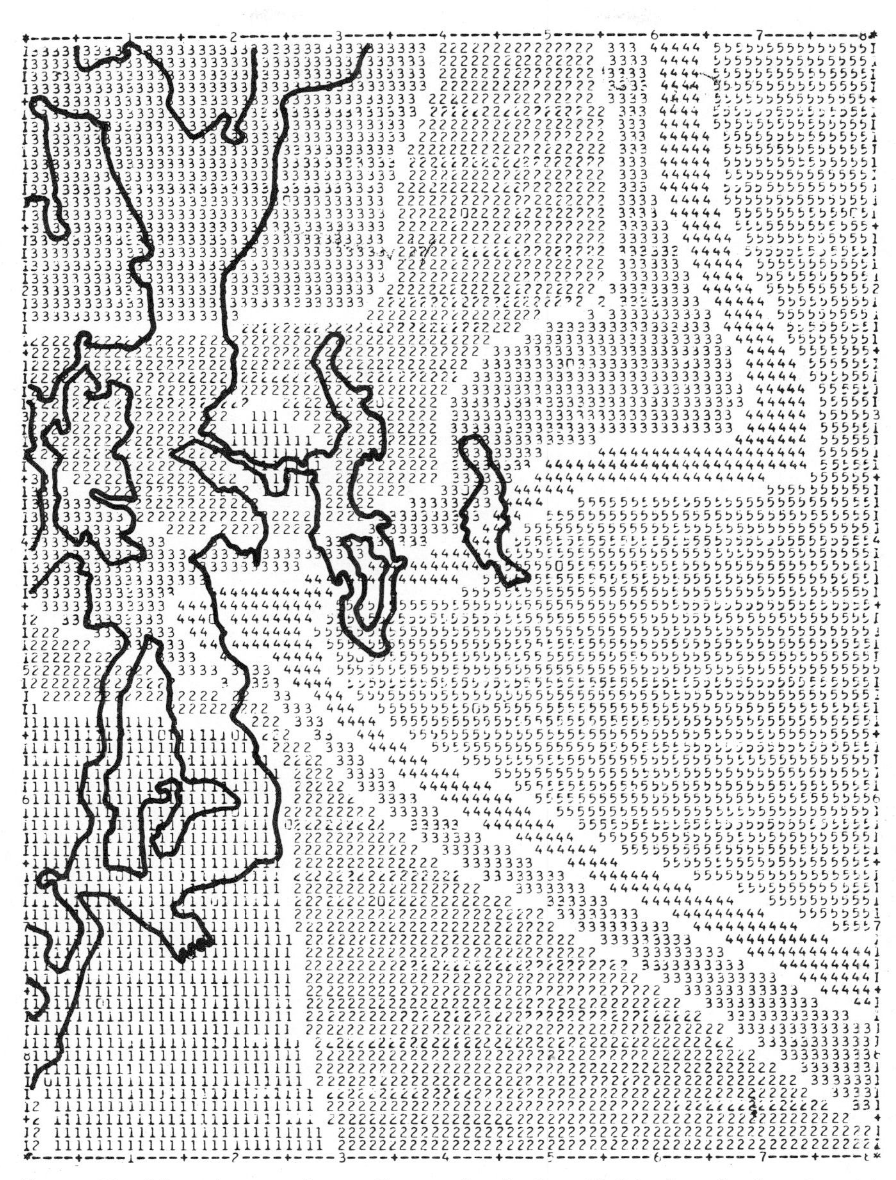

Figure 22. Map of rain volume. Contour levels: 1 = 30–84 ml per bucket; 2 = 84–137 ml per bucket; 3 = 137–191 ml per bucket; 4 = 191–244 ml per bucket; 5 = 244–298 ml per bucket.

Literature Cited

1. C. E. Junge, "Air Chemistry and Radioactivity," Academic Press, New York, 1963.
2. "ARTHUR" is an integrated package of computer programs designed for pattern recognition and factor analysis. It contains, at present, approximately thirty interactive programs. "ARTHUR" was written by D. L. Duewer, J. R. Koskinen, and B. R. Kowalski, and is available from B. R. Kowalski, Laboratory for Chemometrics, Department of Chemistry BG-10, University of Washington, Seattle, Washington 98195.
3. T. V. Larson, R. J. Charlson, E. J. Knudson, G. D. Christian, and H. Harrison, "The Influence of a Sulfur Dioxide Point Source on the Rain Chemistry of a Single Storm in the Puget Sound Region," Water, Air, and Soil Pollution 4, 319 (1975).
4. E. J. Knudson, thesis, University of Washington, 1976.
5. B. R. Kowalski and C. F. Bender, "Pattern Recognition. II. Linear and Nonlinear Methods for Displaying Chemical Data," J. Am. Chem. Soc. 95, 686 (1973).
6. D. L. Duewer, B. R. Kowalski, and J. L. Fasching, "Improving the Reliability of Factor Analysis of Chemical Data by Utilizing the Measured Analytical Uncertainty," Anal. Chem. 48, 2002 (1976).
7. P. Horst, Factor Analysis of Data Matrices, Holt, Rinehart, and Winston, New York, 1965.
8. H. F. Kaiser, "The Varimax Criterion for Analytical Rotation in Factor Analysis," Psychometrika 23, 187 (1958).
9. Y. T. Chien and K. S. Fu, "On the Generalized Karhunen-Loève Expansion," IEEE Trans. Information Theory 13, 518 (1967).

6

Analysis of the Electron Spin Resonance of Spin Labels Using Chemometric Methods

JAMES R. KOSKINEN* and BRUCE R. KOWALSKI

Laboratory for Chemometrics, Department of Chemistry,
University of Washington, Seattle, WA 98195

Spin labels are being used to study the structure of model membrane systems and biological membranes (1,2). The spin labeling technique involves incorporating a nitroxide free radical (the spin label) into a membrane system and studying the free radical using electron spin resonance (ESR) spectrometry. Lipid spin labels that are diffused into a membrane orient themselves in a specific configuration and undergo anisotropic molecular motion. When this motion is rapid on the ESR time scale, the ESR spectra that are observed can be correlated with the structure of the membrane.

Molecules have been constructed so that the long axis of the molecule is parallel to one of the principal axes of the nitroxide. Anisotropic motion about the long axis of the molecule corresponds to rotation about one of the principal axes of the nitroxide. The ESR spectra of this type of molecule in a well defined inclusion crystal have been studied and synthesized in order to better understand the membrane spin labeling experiments (3,4,5).

Studies using spin labels involve a considerable effort for the chemist in the collection and analysis of the spectra. In the past, spectra were collected as two dimensional plots on a piece of paper and the useful information extracted from the plots using a ruler and a pencil. With the introduction of laboratory computers this task has been made much easier (6). Spectra are now collected by computer controlled spectrometers and are saved in computer compatible format (i.e., on paper tape, magnetic tape, or disks). This use of computers also allowed

*Present address: Ford Motor Company
Scientific Research Laboratory, 53061
Box 2053
Dearborn, Michigan 48121

some simple data analysis techniques to be performed on the spectra. These techniques included base-line corrections and spectral smoothing. Computers have made it relatively easy to collect and store all the ESR spectra for a particular study. This paper will present examples of the use of computers to aid the chemist in the analysis of ESR spectra. The first application will involve the use of Chemometric methods to study two spin labels in different inclusion crystals. This application will demonstrate the general usefulness of chemometrics to analyzing ESR spectra. The second application will concern spin labels in a model membrane system.

Methodology

The data analysis methods used in this paper come under the general heading of Chemometrics (7,8). The methods used are the ones that will extract features from the ESR spectra, calculate the importance of the extracted features to a particular property of the spin label, and finally, display the results. All the spectra used in this study were collected under computer control and stored in digital form as 980 data points. The 980 data points can be used as features that describe each spectrum. However, such a large number of features can present difficulties for some data analysis methods. A method that reduces the number of features describing the spectra without losing chemically useful information is clearly needed.

The method of choice in this study is the Fourier transform. Fourier transform methods have been used quite extensively in other forms of spectroscopy for a variety of purposes (9). The effect of the Fourier transform is to condense the information of the total ESR spectrum into the low frequency end of the transformed spectrum. Figure 1 shows a typical ESR spectrum, the real and imaginary parts of the Fourier transform of the spectrum, and the power spectrum. The low frequency end of the transformed spectrum contains all the information needed to reconstruct the original spectrum via the inverse or back transform. This process is graphically presented in Figure 2. The first 64 points of the transformed spectrum are retained while the rest of the points are set to zero. The inverse transform returns the original spectrum showing that no information loss results. The spectrum resulting from the inverse transform appears to be smoother than the original spectrum because the high frequency noise has been digitally filtered by the transform. By using the Fourier transform, 64 features that completely describe the spectrum have been generated out of a spectrum of 980 data points. Once the features have been generated in this manner, the other Chemometric methods can be applied.

Two statistical methods are used to determine the importance of the generated features in modeling a property of the spin label. The property of interest in the first application is the

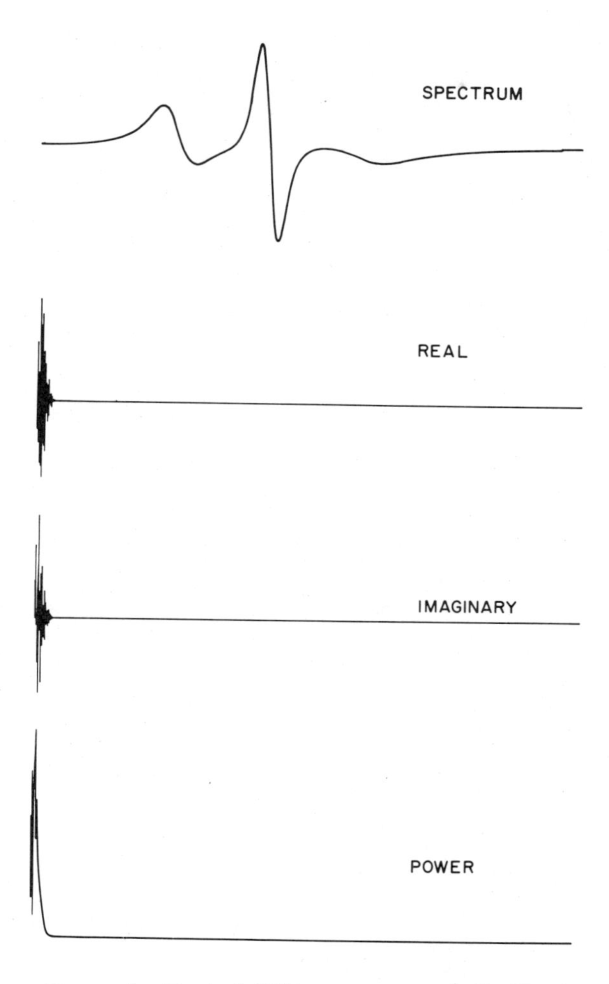

Figure 1. Typical ESR spectrum and its Fourier transform

temperature of the spin label and the property of interest in the second application is the amount of spin label present. The first method calculates the correlation between the generated features and the property. The second method is step-wise regression analysis that determines which of the features does the best job of modeling the property with a linear model. Plots of the generated features vs. the property are also constructed as part of the analysis. All of the methods described are part of the ARTHUR pattern recognition system (10) which was used in this study.

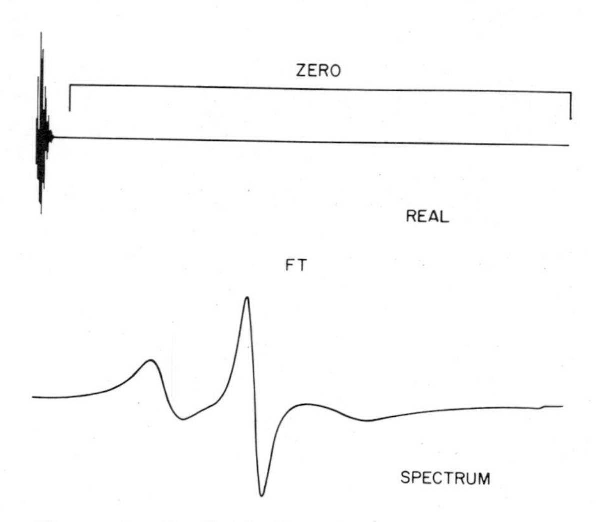

Figure 2. Graphical demonstration that only the first 64 points of the Fourier transform of an ESR spectrum are needed to regenerate the spectrum from the transform

Spin Labels in Inclusion Crystals

The first system studied using the above described methodology consisted of 3-doxyl-5α-cholestane (I) (the 4',4'-dimethyloxazoladine-N-oxyl derivative of 3-keto-5α-cholestane) in an inclusion crystal of thiourea. The question to be answered in this study is: can the temperature of the inclusion crystal system be correlated to the ESR spectrum?

The data set contains 16 ESR spectra of the spin label-inclusion crystal system corresponding to a range of temperatures from $-82.0^{\circ}C$ to $59.2^{\circ}C$. Table I lists the data analysis steps taken to analyze this series of spectra. Feature number four, generated using the Fourier transform, is found to be the most important feature in modeling the temperature of the system. Figure 3 shows a plot of the temperature vs. feature four.

Structure I

Table I
Steps Taken in Data Analysis

I Collect Spectra
II Generate Features Using Fourier Transform
  A Zero Fill Spectra to 1024 Points (Requirement of Fast Fourier Transform)
  B Perform Fast Fourier Transform
  C Select the First 64 Coefficients of the Real Part of the Transform
III Calculate Correlation between the 64 Features and Property
IV Perform Stepwise Regression Analysis of the 64 Features
V Generate Plots of Features Selected in Steps III and IV vs. the Property
VI Analyze Results

Iedally, Figure 3 should show a straight line indicating that feature four is linearly related to the temperature. The scatter of points about the line can be interpreted as meaning that the anistropic motion of the molecule is somewhat restricted. The shorter steps between feature four values at the high temperature end indicates that the rotation about the long axis of the molecule is being optimized.

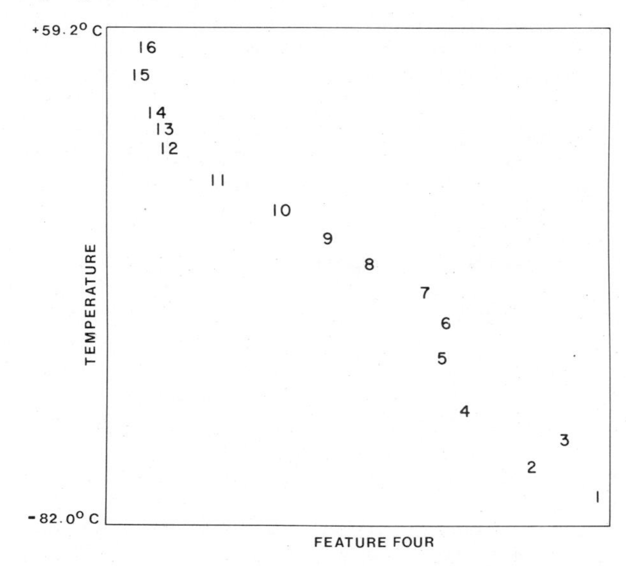

Figure 3. Plot of the Fourier transform generated feature number four (the ordinate) vs. the temperature of the system (the abscissa) in sample one

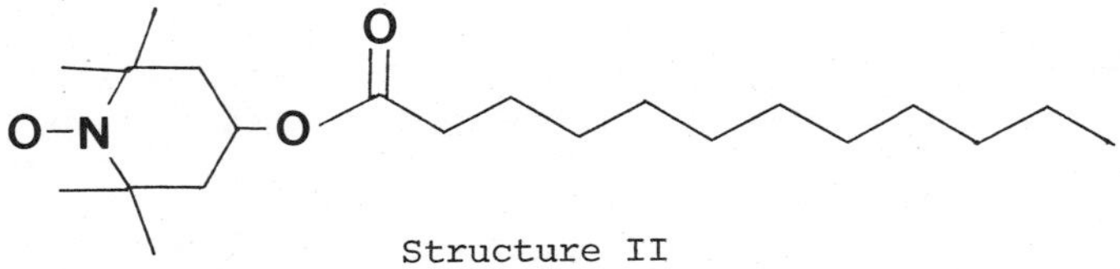

Structure II

The second system studied consisted of the spin label lauryl nitroxide (II) (2,2,6,6-tetramethyl-4-piperidinol-1-oxyl dodecanoate) in an inclusion crystal of β-cyclodextrin. The data set for this system contains 20 ESR spectra of the spin label in the inclusion crystal corresponding to a temperature range of -196°C to 63°C. Table I again lists the data analysis steps taken in the analysis of these spectra.

Fourier transform feature number three is shown by the stepwise regression analysis to be the most important feature in modeling the temperature of the system. Once again a linear plot of the feature and the temperature is expected. Figure 4 shows the actual plot which appears to be linear from the low temperature end (-196°C) to a temperature of about 35°C. Then the value of the feature does not get any larger. It remains nearly constant from about 35°C to 63°C. In the low temperature region, the ESR spectrum approaches the rigid glass limit. As the temperature increases, the molecule starts to rotate more freely about its long axis. At approximately 35°C the rotation about the nitroxide principal x-axis is fast enough on the ESR time scale such that the y and z contributions are averaged out. A further increase in temperature has no additional effect on the anistropic motion.

It is interesting to compare both spin labels in their rigid matrices. The lauryl nitroxide is able to rotate quite freely and reaches an optimum value. The 3-doxyl-5α-cholestane is not able to rotate as freely as the lauryl nitroxide and appears not to reach an optimum value. This difference in rotation can be accounted for by the structure of the molecules. Lauryl nitroxide is a long, cylindrical-shaped molecule, while the 3-doxyl-5α-cholestane is a rectangular shaped molecule. It is easier for the cylindrical molecule to rotate about its long axis in a cavity in a matrix than it is for the rectangular-shaped molecule.

Now that the Chemometric methods have been shown to be useful in the study of spin labels in well-defined inclusion crystals the methods can be used in the study of spin labels in a model membrane system. The last part of this paper will deal with the application of Chemometric methods to the study of such a model membrane system.

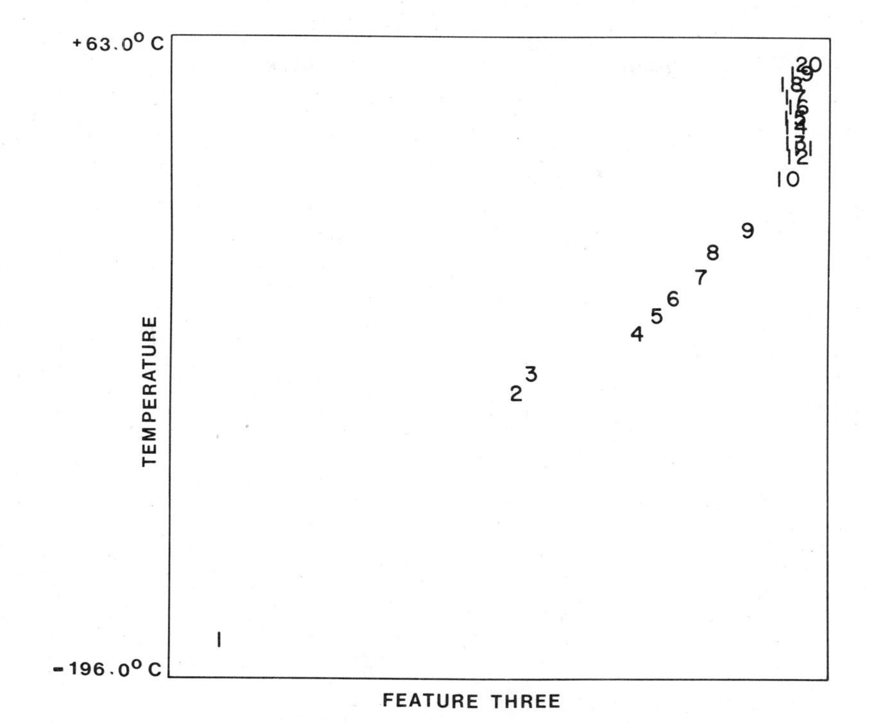

Figure 4. Plot of the Fourier transform generated feature number three (the ordinate) vs. the temperature of the system (the abscissa) in sample two

Spin Labels in Model Membrane Systems

The model membrane system studied is the cytochrome oxidase protein containing spin labeled phospholipids. The spin label used is 16-doxyl steric acid (III) (the 4',4'-dimethyloxazoladine-N-oxyl derivative of 16-keto stearic acid). Figure 5 shows the spectra of representative samples of the cytochrome oxidase protein with different concentrations of phospholipids. The amount of lipid in each sample is expressed as the ratio of mg of phospholipid per mg of protein. The sample in Figure 5a has a ratio

Structure III

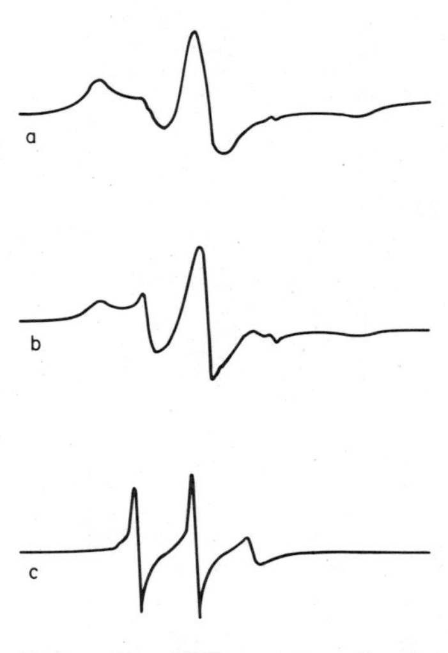

Figure 5. ESR spectra of spin labeled phospholipids in cytochrome oxidase (see *text)*

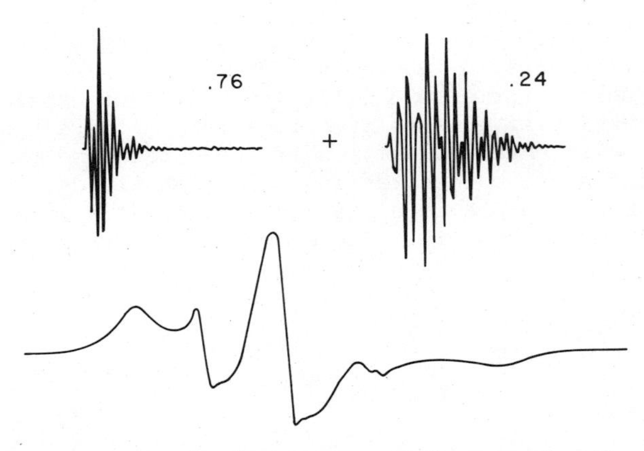

Figure 6. Graphical presentation of the generation of a composite ESR spectrum by using scaled amounts of Fourier transform of two ESR spectra

of 0.10; Figure 5b corresponds to a ratio of 0.24; and Figure 5c is 0.73. The ESR spectrum of the sample with the lowest lipid content (Figure 5a) is characteristic of strong immobilization of the spin labels while the spectrum of the sample with the highest lipid content (Figure 5c) is characteristic of a more mobil spin label (11). The question to be answered in this experiment is: is it possible to quantify the amount of each kind of spin label in a composite system as shown in Figure 5b?

The data set includes eight spectra of samples of varying amounts of the protein and spin labeled phospholipid. The feature generation methodology used is the same as described in the previous examples. The property of interest in this example is the amount of immobilized lipid present in the model membrane system. By using stepwise regression analysis it is possible to arrive at an equation to calculate the amount of the lipid present.

By using this equation, it is possible to look at the Fourier transform of an ESR spectrum of the membrane system and calculate the amount of immobilized lipid present. A practical proof of the validity of this equation is to synthesize an ESR spectrum for the series studied using spectra of the immobilized spin label and the mobil spin label. Since the equation was developed using the Fourier transform to the ESR spectra, they will be used in place of the spectra. The Fourier transform of the immobilized spin label spectrum (Figure 5a) is multiplied by the calculated scale factor and the result is added to the Fourier transform of the mobil spin label scaled by the calculated factor. Then the inverse transform is applied to this composite to give the spectrum. In this case the Figure 5b is the spectrum that is being synthesized. The scale factor for the immobilized spin label is 0.24, and the factor for the mobil spin label is 0.76. This process is shown graphically in Figure 6. The resultant synthetic spectrum appears smoother than the experimental spectrum because the high frequency noise has been digitally filtered.

In this application Chemometric methods were used to show that certain ESR spectra of a model membrane system are a composite of spectra of an immobilized spin label and a mobil spin label.

Conclusions

Chemometric methods have been used to analyze experimental ESR spectra. The methods have provided additional insight into the processes involved in putting a spin label into an inclusion crystal. They have also been used to examine the ESR spectra resulting from spin labels dispersed in a model membrane system. Chemometrics does provide a powerful tool to aid the chemist in the analysis of ESR spectra of spin labels in model membrane systems.

Acknowledgments

We wish to acknowledge Drs. Patricia Jost and O. Hayes Griffith for kindly providing us with the ESR spectra used in this study. We also wish to acknowledge the financial support of the Office of Naval Research under Contract No. N00014-75-C-0536.

Literature Cited

1. McConnell, H. M. and McFarland, B. F., Quart. Rev. Biophys., (1970) 3, 91-136.
2. Jost, P., Waggoner, A. S., and Griffith, O. H., in "Structure and Function of Biological Membranes," Rothfield. L. (Ed.) pp. 84-144, Academic Press, New York, 1971.
3. Birrell, G. B., Van, S. P., and Griffith, O. H., J. Amer. Chem. Soc., (1973) 95, 2451.
4. Birrell, G. B., Griffith, O. H., and French, D., J. Amer. Chem. Soc., (1973) 95, 8171.
5. Griffith, O. H., J. Chem. Phys., (1964) 41, 1093.
6. Klopfenstein, C. E., Jost, P., and Griffith, O. H., in "Computers in Chemical and Biochemical Research," Klopfenstein C. E. and Wilkins, C. L. (Eds.), pp. 175-221, Academic Press, New York, 1972.
7. Kowalski, B. R., Anal. Chem. (1975), 47, 1152A.
8. Kowalski, B. R., J. Chem. Infor. & Compt. Sci., (1975), 15, 201.
9. Marshall, A. G. and Comisarow, M. B., Anal. Chem., (1975), 47, 491A.
10. Duewer, D. L., Koskinen, J. R. and Kowalski, B. R., "ARTHUR" available from B. R. Kowalski, Laboratory for Chemometrics, Department of Chemistry, University of Washington, Seattle, Washington 98195.
11. Jost, P. C., Griffith, O. H., Capaldi, R., and Vanderkooi, G., Proc. Nat. Acad. Sci. USA, (1973), 70 (2), 480-484.

7

Automatic Elucidation of Reaction Mechanisms in Stirred-Pool Controlled-Potential Chronocoulometry

LOUIS MEITES and GEORGE A. SHIA

Department of Chemistry, Clarkson College of Technology, Potsdam, NY 13676

During a symposium that deals with techniques for processing chemical data, it is appropriate to inquire why such techniques are important, and what effects their adoption may eventually have on the chemist's work and thought.

They are important for several different reasons. They can facilitate calculations and interpretations that may involve many steps and, with numerical data, tedious graphical or other analysis. By doing so they can save much of the chemist's time and energy. At the same time they can yield more reliable results because they substitute the dumb patience and objectivity of a machine for the human frailty and occasional unconscious prejudice of the chemist. This makes it possible to find correlations or interpretations that the chemist might miss, and to achieve greater depth and certainty in the final result. They can influence the design of experiments in several ways: by making it possible to find a data-acquisition schedule that stresses the regions of greatest importance to the desired result and enables the experimenter to ignore others of lesser importance, by making it possible to obtain the desired result from a simple experiment and thus obviating the necessity of performing a more complicated one that would yield the same information in a form more easily amenable to older techniques of data analysis, and even by employing the data so efficiently that one experiment can be made to yield certainty as great as could have been obtained from three or four with the aid of the older techniques.

Examples of all of these are already in the literature, and new ones continue to appear. It is already evident that the growing adoption of these techniques is substantially easing the tedium of experimentation and interpretation while improving the accuracy and reliability of the values deduced from the data. As it relieves chemists of unnecessary burdens, it will have longer-range effects on chemical education, which now of necessity devotes time to the teaching and learning of techniques of data analysis that might better be spent on such things as the design of experiments and the interpretation of results.

In the course of making these potentialities manifest, some bastions of impossibility have already crumbled, as may be shown by several examples from the field of potentiometry and potentiometric titration. It has always been known that analyses by direct potentiometry cannot be made as accurately and precisely as analyses by potentiometric titration, but Brand and Rechnitz (1) and Isbell, Pecsok, Davies, and Purnell (2) have shown that this is not so. It has always been known that acetic acid (pK_a = 4.755) and propionic acid (pK_a = 4.876) have strengths too nearly identical to permit identifying, much less determining, both from the acid-base titration curve for a mixture, but Ingman *et al.* (3) have shown that this is not so. It has always been known that a potentiometric acid-base titration cannot succeed if there is no point of maximum slope on the titration curve or if the concentration of the reagent is unknown, but Barry and Meites (4) and Barry, Campbell, and Meites (5) have shown that these are not so. These are only a few of the many instances in which it is now apparent that familiar experiments have always provided us with information that we have not known how to obtain in useful form.

Probably there are very few chemists who would not readily concede the effectiveness of computerized procedures in evaluating numerical parameters on the basis of numerical data. Applications and examples like the ones just cited are therefore comparatively easy to understand and accept. However, there are many fewer chemists who are prepared to accept the idea that accurate and reliable qualitative decisions can be made by machines without human intervention. In human terms it is of course understandable why this should be so. The chemist faced with having to decide whether a compound contains, say, a phenyl group on the basis of its mass spectrum is sure to be aware, at some level, of the years of training and experience that he brings to that decision, and of all the subtleties and pitfalls that it may involve. It is no easy thing to admit that one's knowledge, understanding, and insight cannot produce decisions superior to those made by a mindless machine - or, more properly, that that knowledge, understanding, and insight can be reduced to a set of completely predetermined steps that will produce as good a decision as the one that could be made by a human brain.

Despite this difficulty, it is clear that progress in this area offers prospects having overwhelming importance. Though the portion of chemical research that deals with the evaluation of numerical parameters can certainly be significant and challenging, the portion that deals with qualitative interpretation is even more so. In experimental kinetics, for example, the real problem is usually to decide what the mechanism of a reaction is. After this has been solved, more or less severe experimental difficulties may still have to be overcome in evaluating the rate and equilibrium constants associated with the reaction, but often the values of these will be implicit in the data accumulated in elucidating the mechanism.

What is involved in making such decisions? There are two possibilities. One is that the system being studied belongs to one of a limited number of classes whose behaviors are accurately known. The other is that it does not, and instead behaves in some novel way that no prior interpretation will suffice to describe.

In the first case the interpretation may be said to be routine. Of course the decisions that are involved in assigning the system to the proper class may be both subtle and difficult. Calling them routine merely means that similar decisions have been made before for other systems, that the principles underlying those decisions are known, and that it should therefore be possible to effect them in ways that a computer can be programmed to execute. In the second case only enough is known to make it possible to decide that the system does not belong to any known class. Beyond that point decision and interpretation require imaginativeness and intuition, and these cannot be programmed in advance. Except in situations so simple that all the possible classes are already known, every program designed to make such interpretations must provide for a call for human intervention when all of the known possibilities have been tested and found to be inadequate.

The nature of research in any area would be profoundly changed by the availability of a program that would effect routine interpretation and that would call for human help when this did not suffice. Chemists working in that area would be relieved of the necessity of undertaking such interpretation themselves - of retracing on each problem the thoughts they had had while solving the one before it. In losing this burden they would gain the opportunity to spend more of their time in breaking new ground, in improving the excellence of their experimental procedures and measurements, and in selecting the systems that would best repay study. Investigations that turned out to be routine would be greatly facilitated, and those that did not could receive the benefit of the human imagination thus liberated.

This is the rationale of a program of research that began several years ago at Clarkson College of Technology. Near its start we identified the various kinds of classifications that might arise (<u>6</u>), and these are listed in Table 1.

Table 1. Kinds of Classifications

1. Binary
    - A. Simple
    - B. Multiple
2. Linear
3. Branched

It is always supposed that data are available showing how the value of some dependent variable changes as that of some independent one is altered, and that the problem is to account for the data by selecting the appropriate one of a finite set of hypotheses, each of which can be expressed by one or more equations relating the dependent and independent variables and involving numerical parameters as well.

A simple binary classification is one that requires only a single "yes-no" decision. A compound may or may not contain a particular kind or group of atoms; it may or may not have a particular kind of biological activity; if subjected to polarographic examination it may or may not be reversibly reduced. Much of the research to date on pattern recognition in chemistry has dealt with such questions. It may or may not be possible to express one or both of the alternatives by an equation relating the independent and dependent variables: our work has concentrated on cases in which this is possible, while pattern recognition has concentrated on those in which it is not. All such questions share the property that the two alternatives being considered are both exhaustive and mutually exclusive: if either can be rejected the other must be right. For the kinds of data considered here, there may be two different possible equations of which the data must conform to one, or there may be only one equation and the question may be whether the data conform to that equation or not. In either event it is relatively simple to express the reliability of the classification in the classical statistical fashion: that is, by stating the level of confidence at which it can be upheld.

Multiple binary classifications are those that require two or more independent "yes-no" decisions. A not very complicated one is shown in Fig. 1. This arose in potentiometric titrations of sodium or potassium laurate with hydrochloric acid (7). Depending on the concentrations of laurate and hydrogen ions in these two solutions, and also on the concentration of any alkali-metal salt (such as sodium or potassium chloride) that was added, any of three different and independent phases might have separated during some portion of the titration. Micelles of laurate ion might or might not have been present in the initial solution, and both an "acid soap" and the free fatty acid might or might not have precipitated. Eight different kinds of titration curves can result, although not all can be obtained with laurate and sodium or potassium ions in aqueous solutions at 25°. A program was constructed to effect this triple binary classification, beginning with exactly the same information - the compositions of the reacting solutions and the coordinates of the points on the experimental titration curve - that would be available to a human chemist. When it was applied to about 30 experimental curves secured under widely varying conditions, its error rate was approximately 5%. Two skilled chemists using the same information managed, with some difficulty, to achieve an error rate of about 35%.

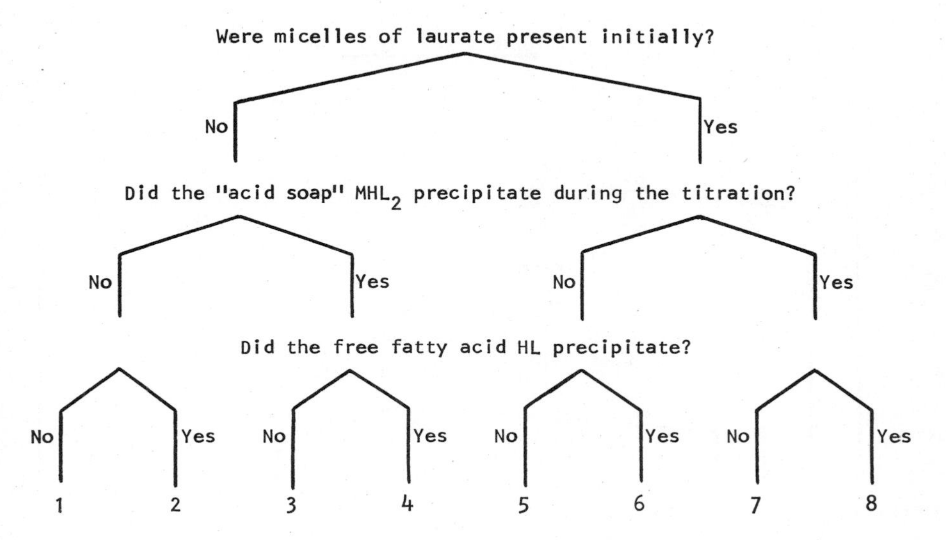

Figure 1. Triple binary classification of the potentiometric titration curve obtained in a titration of laurate ion with a strong acid. The purpose of the classification is to reveal which, if any, separate phases were present at any stage of the titration.

Linear classifications may be defined as those in which the equations that correspond to the successive hypotheses can be arranged a priori in a logical order that is also the order of increasing complexity of the equations. Figure 2 shows the simplest example in the literature (6). In some cases the number of possible hypotheses may be very large, but in others, including this one, it may be very small. The literature contains no example of any tetrafunctional base whose dissociation constants are all so close together that successive steps could not be perceived on visual inspection of the titration curve, and therefore it was decided that no hypothesis beyond the tetrafunctional one would be allowed. Together with the fact that the successive decisions were based on an estimated standard deviation of measurement, this decision made it possible to close the loop and ensure the eventual acceptance of one of the permissible hypotheses.

Programs that effect linear classifications are easier to design than those for multiple binary classifications, and are

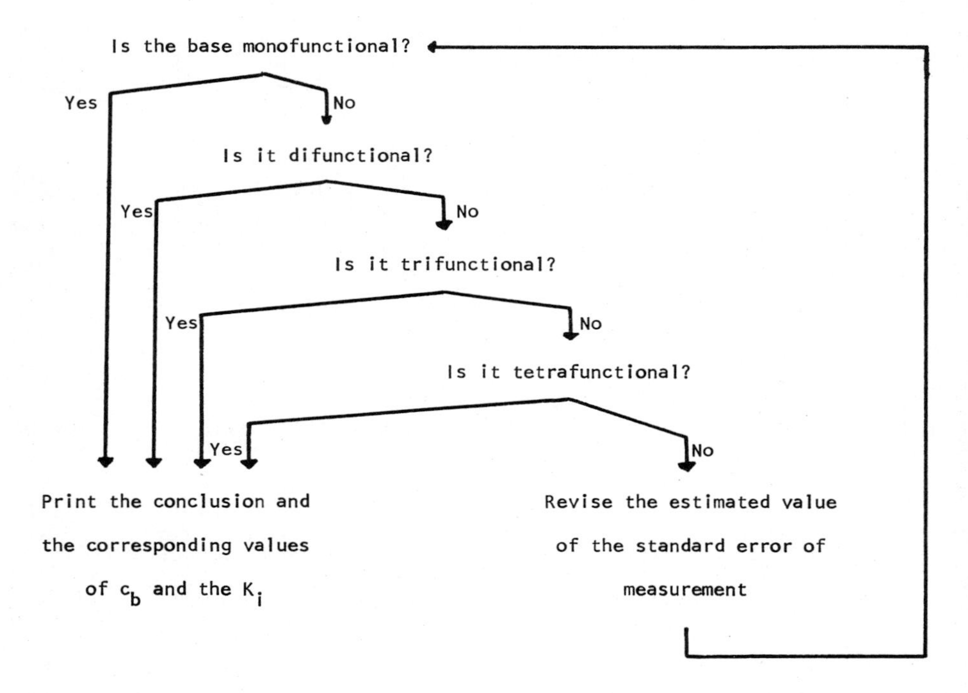

Figure 2. Linear classification of the potentiometric titration curve obtained in a titration of a weak base with a strong acid. It is assumed that no phase separation occurs during the titration. The purpose of the classification is to reveal the number of protons consumed by each ion or molecule of the base during the titration.

likely to be at least as reliable, if not more so. The one represented by Fig. 2 was tested with a great many synthetic and experimental data for bases like acetate, succinate, and citrate, and could not be made to fail when it was provided with an honest estimate of the standard error of measurement. If it was given an estimate that was much too large, it wielded Occam's Razor with ruthless skill, exactly as a human chemist would, accepting the simplest hypothesis that could not be disproved. Occasionally, when it was given an estimate that was nearly a full order of magnitude too small, it did conclude that citrate ion was tetrafunctional; human chemists are often similarly misled by data that are less precise than they are thought to be.

Very much the most complicated kind of classification is the branched classification, in which the rejection of one hypothesis must be followed by a test of another one, and in which there is no logically necessary order in which the tests must be arranged. This paper describes the structure of the first program designed to effect a branched classification.

The Chemical Problem

Controlled-potential electrolysis has been studied intensively in several laboratories since about 1955 with a view to employing it for the elucidation of the mechanisms of electrochemical processes and for the evaluation of the rate and equilibrium constants for the individual steps in these processes. Roughly 25 different mechanisms have been imagined. A few were invented *a priori* or studied because they were known to have consequences that could be observed by other electrochemical techniques; most were devised in order to account for the phenomena observed in studying real systems, both organic and inorganic; a few were investigated briefly so that they could be ruled out as being unable to explain those phenomena. There are several thorough reviews (8-10) stressing diagnostic criteria and ways of differentiating among the various mechanisms, and only a very brief summary will be undertaken here.

Controlled-potential electrolyses are performed in cells fitted with three electrodes. One, the "working electrode," is the electrode at which the half-reaction of interest occurs. Often it is a large pool of mercury, though platinum gauze and other materials can also be used. The solution surrounding the working electrode is efficiently stirred to facilitate mass transfer of the reactant from the bulk of the solution to the surface electrode and to maintain a constant value of the mass-transfer coefficient s. An applied potential is imposed across the working electrode and an "auxiliary electrode," which is usually isolated in a separate compartment of the cell and serves merely to permit the flow of an electric current through the cell and working electrode. The zero-current potential $E_{w.e.}$ of the working electrode is sensed by comparing it with the potential of a "reference electrode," such as a saturated calomel or silver-silver chloride electrode, through which current does not flow. Altering the applied potential causes the value of $E_{w.e.}$ to vary. By means of a potentiostat, the applied potential is continuously adjusted to as to keep $E_{w.e.}$ equal to some predetermined value, which is usually so chosen that ions or molecules of the reactant A are reduced as rapidly as they are brought to the surface of the working electrode by convection and diffusion. Of course oxidation is also possible, but only reduction will be alluded to here.

The simplest possibility is that A is reduced to a stable product P in a single step and without any side reaction or other coupled chemical process. This is described by the equations (11)

$$A + n\,e + P \tag{1}$$

$$i = i^{o}e^{-s\,t} \tag{2a}$$

$$Q = Q_{\infty}\,(1 - e^{-s\,t}) \tag{2b}$$

where n is the number of faradays consumed in reducing each mole of A, i is the current that flows through the cell and working electrode t s (seconds) after the electrolysis has begun, Q is the quantity of electricity (coulombs or faradays) that has flowed up to that instant, i^{o} is the initial value of i, and Q_{∞} is the total quantity of electricity that will flow if the electrolysis is indefinitely prolonged. The value of Q_{∞} in faradays is given in this simple case by Faraday's law:

$$Q_{\infty} = n\,N^{o}_{A} \tag{3}$$

where N^{o}_{A} is the number of moles of A present in the initial solution.

Different mechanisms yield different results. It is possible for A to undergo reductions along parallel but independent paths to yield different products;

$$A + n_{1}e = P \quad ; \quad A + n_{2}e = Q \tag{4}$$

Then the current will decay exponentially with time, as it does in the simple case, but the apparent value of n defined by the equation

$$n_{app.} = Q_{\infty}/N^{o}_{A} \tag{5}$$

will lie between the values of n_{1} and n_{2} and will reflect the relative magnitudes of the overall rate constants for the two half-reactions. There may be a "continuous faradaic background current" $i_{f,c}$ (12) due to the reduction of hydrogen ion, water, or some other major constituent of the solution, at a rate that is constant and unaffected by the concentration of A or P. Then the current is given by

$$i = (i^{o} - i_{f,c})e^{-s\,t} + i_{f,c} \tag{6}$$

where i^{o} is the total current at $t = 0$. The quantity of electricity is given by

$$Q = Q_{\infty}\,(1 - e^{-s\,t}) + i_{f,c}t \tag{7}$$

Behavior very similar to this results from the mechanism

$$\begin{aligned} A + \underline{n}\,\underline{e} &= P \\ P + Z &= A + Z \end{aligned} \qquad (8)$$

where Z is some major constituent of the solution and Z' is its reduced form. Frequently Z is hydrogen ion and Z' is hydrogen gas. Again the current decays to a finite steady-state value, which depends on the rate constant for the chemical step in this mechanism and on the concentration of Z. At any instant the "regeneration current" is equal to the difference between the current that is actually observed and the one that would be observed if the chemical step did not occur. In contrast to a continuous faradaic background current, which has the same value thoughout an electrolysis, the regeneration current is equal to zero at the start and increases toward the steady-state value as the electrolysis proceeds and the concentration of P increases. Only by experimental measurements far more precise than any ever made could it be detected that equations (6) and (7) provided slightly imperfect fits, but because equation (7) overestimates the quantity of electricity consumed by regenerated A near the beginning of the electrolysis it yields a value of $\underline{Q}_{\infty}$ that is slightly smaller than the one described by equation (3).

For some pairs of mechanisms the differences of behavior are prominent and unmistakable; for others they are subtle and difficult to detect. The aims of the present work are

1. to devise a program that will identify the mechanism of a process, using no more information than would be available to a research chemist studying the same process for the first time;

2. to study the nature and reliability of the classifications that such a program can make and, in doing so, to strive for additional insight into the thought processes of the human chemist attempting to make similar classifications; and

3. to establish some general principles and expectations that may facilitate the development of other branched classification programs in the future.

The Starting Point

The first necessity was that of deciding what data and information might be provided to the program. Either the current or the quantity of electricity might be measured as a function of time. Especially when the working electrode is a violently stirred mercury pool, it has long been known that the noise level of the measured current is rather high, typically of the order of $\pm$ 10 per cent. This is due partly to transient fluctuations of the area of the pool, and partly to the inherent irreproducibility of the convective mass-transfer process. These momentary irregularities are smoothed out by current integration to such an extent that even those employing equations describing the dependence of current on time have usually chosen to evaluate the current at time t by employing the equation

$$i_t = \frac{Q_{t+\Delta t} - Q_{t-\Delta t}}{2(\Delta t)} \qquad (9)$$

in preference to measuring that current directly. Direct-reading electromechanical current integrators with precisions as good as 0.1 per cent of full scale, or better, have been available for over 20 years, and can now be obtained with BCD outputs for direct data acquisition. Experimentation with measurements of current and current integrals convinced us that very much better fits to theoretical equations could be obtained for the latter. Consequently we decided to assume that Q-t data would be available and that the absolute error in a measured value of Q would be independent of Q.

In addition to the coordinates of a number of points on the Q-t curve, the chemist trying to elucidate the mechanism of a process would surely know several things about the experimental conditions, and this information is obtained in a short initial interactive dialogue. It may be said here that the program is written in the Digital Equipment Corporation's EDUSystem 25 BASIC for execution on a PDP8/l minicomputer that provided 4096 words of core memory as the user area available for this work. Partly because of the number of hypotheses that must be tested, and partly because lengthy print commands are included to provide detailed guidance in designing subsequent experiments, the entire length of the program considerably exceeds 4096 words. It is therefore constructed in several segments chained together, with data files on magnetic tape for transferring numerical values from one segment to another.

It is the first segment that contains the data statements and the initial dialogue. Typical examples of these are shown in Figs. 3 and 4. The data statements are arranged in the form t_1, Q_1, t_2, Q_2,... . The initial dialogue obtains the number of millimoles of starting material in the original solution, the volume of that solution, and the expected value of n. Because we thought it important to enable the program to accept the simplest hypothesis consistent with the precision of the data, as a human chemist would naturally do, the user's estimate of the standard error in the measured values of Q is obtained as well.

```
700 DAT 10,.0957, 20,.1805, 30,.2605, 40,.3291, 50,.3947
705 DAT 60,.4518, 70,.5041, 80,.5505, 90,.5932, 120,.6972
710 DAT 150,.7755, 180,.8371, 210,.8791, 240,.9099, 270,.9320
715 DAT 300,.9495, 360,.9721, 420,.9849, 480,.9916, 540,.9952
720 DAT 600,.9972, 660,.9980, 720,1.0002, 780,.9992, 840,.9998
725 DAT 900,1.0007
```

Figure 3. Typical data statements

```
RUN

A=STARTING MATERIAL; B=ELECTROACTIVE IMPURITY; I,J,K=IN-
   TERMEDIATES; P,Q,R=PRODUCTS; Y=ELECTROINACTIVE PRECURSOR
   OF A; Z=AN ELECTROINACTIVE SPECIES, OFTEN H-ION, WHOSE
   CONCENTRATION REMAINS CONSTANT DURING THE ELECTROLYSIS;
   Z'=THE REDUCED FORM OF Z

MMOLES OF A TAKEN? 1.0014
EXPECTED N-VALUE? 1
CM^3 OF SOLN.? 52.0
NO. OF DATA POINTS? 26
UNITS OF Q(MEAS.): 0=COULOMBS, 1=MF? 1
ESTD.STD.DEV. OF Q(MEAS.)? 1E-3
IS I(F,C) KNOWN FOR THE BLANK SOLN.: 0=NO, 1=YES? 0
DIAGNOSTIC PRINT WANTED: 0=NO, 1=YES? 1
  HOW MUCH: 1=FINAL, 2=EVERY PTH CYCLE, 3=ALL? 2
      P=? 10
```

Figure 4. Initial interactive dialogue for the data shown in Figure 2

The program reads the data, converts the values of $\underline{Q}$ to millifaradays if they were provided in coulombs, and obtains a value of $\underline{i}_{f,c}$ for the blank solution if this is known. There is no human intervention after this point.

The Body of the Program

Most current-time curves (on which certain types of behavior are easier to discern than they are on $\underline{Q}$-$\underline{t}$ curves) have the general shape shown in Fig. 5: the current decays smoothly and monotonically from its initial value to a very much smaller one which, as was mentioned above, may or may not be indistinguishable from zero. Curves of this sort belong to what is called the "main line" below.

Two different kinds of curves are shown in Figs. 6 and 7. The one in Fig. 6 results from a mechanism like

$$\begin{aligned} &A + 2e = P \qquad \text{(slow)} \\ &P + A \longrightarrow 2I \qquad\qquad (10) \\ &I + e = P \qquad \text{(fast)} \end{aligned}$$

which has been observed in reductions of vanadium(IV) (=A) to vanadium(II) (=P): the starting material and product react in the bulk of the solution to produce vanadium(III) (=I), and one of the curiosities of the electrochemical behavior of vanadium is that the reduction of vanadium(III) is often much more nearly reversible than that of vanadium(IV).

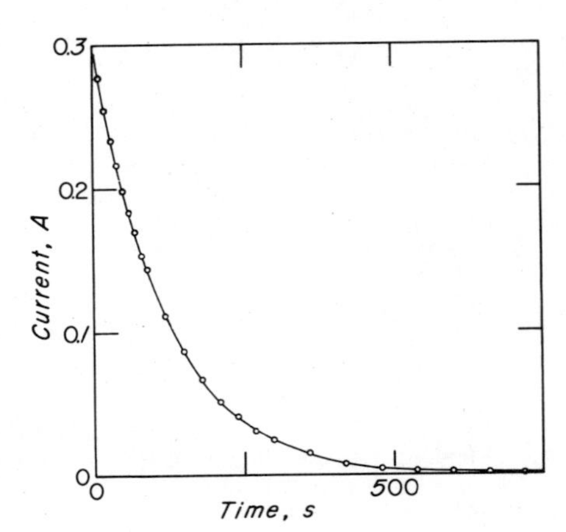

Figure 5. Current–time curve for the controlled-potential reduction of cadmium(II) in 3M hydrochloric acid at a large stirred-mercury-pool electrode maintained at a potential of −0.80 V vs. S.C.E.

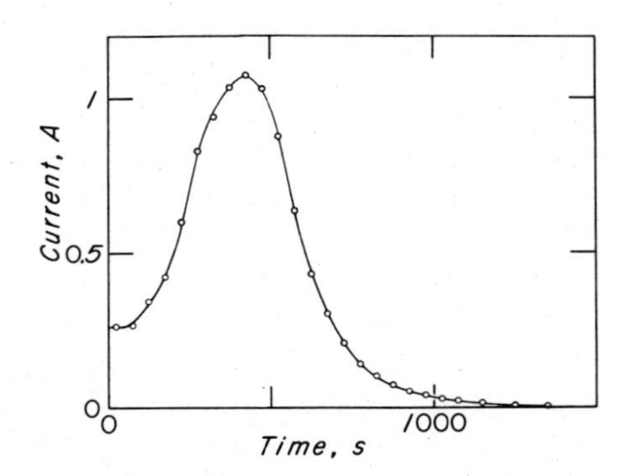

Figure 6. Current–time curve for the controlled-potential reduction of vanadium(IV) in 3M hydrochloric acid at a large stirred mercury-pool electrode maintained at a potential of −0.85 V vs. S.C.E.

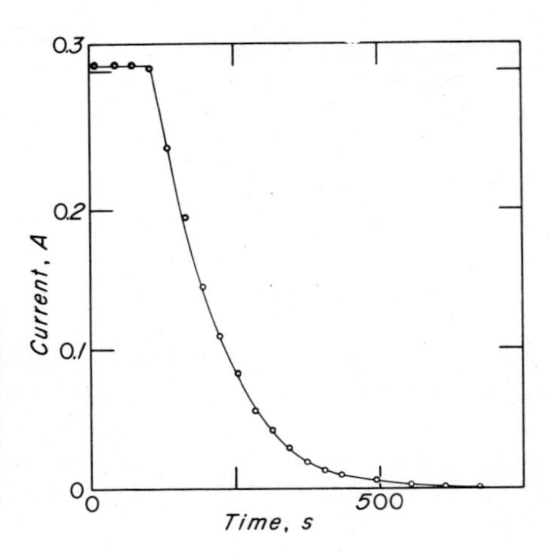

Figure 7. Current–time curve for the controlled-potential oxidation of vanadium(II) in 3.5M hydrochloric acid at a large stirred mercury-pool electrode, with a resistor connected in series with the cell to limit the current at the start of the electrolysis. After the potentiostat gained control the potential of the working electrode was maintained at −0.35 V vs. S.C.E.

The successive values of Q and t are differentiated numerically to obtain values of i:

$$i = \frac{Q_{t_{j+1}} - Q_{t_j}}{t_{j+1} - t_j} \tag{11}$$

The value of i for the first interval (j=1) is stored, and another is computed for the interval at whose end Q first exceeds one-half of its value at the very last point. If the latter current considerably exceeds the first one, it is concluded that there is a maximum on the current-time curve. Since the mechanism described by equations (10) is the only one yet devised that will account for a maximum, crude estimates of the appropriate parameters are derived from the data and stored in a data file, and a second segment of the program is called into core to effect a fit to the differential equation corresponding to that mechanism. An acceptable fit may or may not be obtained. If one is obtained, the values of the parameters are printed out, together with a message saying that only this one of the mechanisms in the program's library is capable of yielding a satisfactory fit. If an acceptable fit is not obtained, a message to that effect is printed out, along with a statement that human interpretation is essential. In either event execution stops at this point.

If the i-t curve does not possess a maximum, the data are next searched for the characteristic features of Fig. 7. This is done by comparing the quantities of electricity accumulated during the first two intervals. If

$$Q_{t_2} > \frac{t_2}{t_1} Q_{t_1} - 3s_Q \tag{12}$$

where Q_{t_j} is the total quantity of electricity accumulated between $t=0$ and $t=t_j$ and s_Q is the estimated standard error of a single measurement of Q, it is concluded that the current-time curve is initially flat within the precision of the measurements. This sort of behavior has been observed in either of two circumstances:

1. when the initial current is controlled by an experimental artefact, such as
   a. a cell resistance so high that the maximum applied potential available from the potentiostat does not suffice to drive $E_{w.e.}$ to the desired value, or
   b. the presence of a separate solid or liquid phase suspended in the solution being electrolyzed, and replenishing the electroactive material dissolved in that solution as rapidly as it is removed by

reduction.

2. when the process follows an ECE mechanism

$$A + n_1e = J$$

$$I \longrightarrow J \qquad (13)$$

$$J + n_2e = P$$

and the rate constant for the intervening chemical step is about an order of magnitude larger than the overall rate constants for the electron-transfer steps.

The first of these corresponds to the equations

$$Q = Q_{t*}(t/t^*) \qquad (t < t^*)$$
$$Q = Q_{t*} + Q'[1-\exp\langle -s(t-t^*)\rangle] \qquad (t > t^*) \qquad (14)$$

where t^* is the time at which the current just ceases to be constant, Q_{t*} is the quantity of electricity accumulated up to that time and Q' is the quantity of electricity that remains to be accumulated thereafter, and s is the overall rate constant for the reduction. There appear to be four parameters, but since it can be shown that $Q' = Q_{t*}/ s\ t^*$ there are only three that are independent. Initial estimates of these three are made and control is transferred to a third segment of the program to effect a fit to these discontinuous equations.

If an acceptable fit is not secured, the ECE mechanism is hypothesized and a fit is made to the differential equation that describes it. If this also does not yield an acceptable fit, the experimenter is advised to repeat the experiment under suitably modified conditions, for the failure of both these hypotheses must mean that an experimental artefact is superimposed on some other mechanism included in the main line, and this makes further analysis injudicious.

We turn now to the main line, which includes current-time curves that do not have maxima and that are concave upward from the very beginning of the electrolysis. The simplest hypothesis that can account for such a curve is the one described by equation (2b):

$$Q = Q_\infty(1-e^{-s\ t}) \qquad (2b)$$

The basic program for effecting non-linear regression requires initial estimates of the values of Q_∞ and s, and these are derived from the raw data, The value of Q_∞ is estimated to be equal to that of Q at the last experimental point. The data are then scanned to find the first point at which Q/Q_∞ exceeds 0.865 ($= 1-e^{-2}$), and the time t' at this point is combined with the equation

$$\underline{s} = s/\underline{t}' \qquad (15)$$

to obtain an estimate of $\underline{s}$.

Regression onto equation (2b) is then effected, using the latest version of a general program (12) that was written locally and has been used for a large number of different purposes in our laboratories and elsewhere. When the fit is complete, the best values of the parameters are combined with equation (2b) to compute a value of $Q_{calc.}$ for comparison with the measured one $Q_{meas.}$. If the difference $(Q_{meas.} - Q_{calc.})$ were plotted aganist the independent variable t the result would be a "deviation plot." Earlier papers from this laboratory (12-15) have demonstrated the utility of deviation plots in testing hypotheses and guiding the selection of better ones, and since this is the basis of the decision mechanism in the present program a brief review of the principles underlying their use is given in the paragraphs that follow.

A deviation plot is a plot of $(y_{meas.} - y_{calc.})$ against x, where x is the independent variable and y the dependent one, $y_{meas.}$ is a measured value of y, and $y_{calc.}$ is the corresponding value of y calculated from an assumed relation between x and y with the numerical values of the parameters that yield the best fit to that relation.

On the assumption that the experiment has been properly designed and is therefore free from systematic errors of measurement, the only errors that can arise if the assumed relation is correct will be random errors. The deviation plot will then consist of points randomly scattered around the x-axis.

If, however, the assumed relation is incorrect, its incorrectness represents a source of systematic error, and the deviations will no longer be perfectly random. They will instead scatter around a curve having a characteristic shape, which depends only on the natures of the equations that describe the assumed and true relations between x and y.

For example, Figure 8 shows the characteristic shape that results from assuming that the relation between Q and t is

$$Q = Q_{\infty}(1-e^{-\underline{s}\,t}) \qquad (2b)$$

when it is actually

$$Q = Q_{\infty,corr}(1-e^{-\underline{s}\,t}) + i_{f,c}t \qquad (7)$$

This is a smooth curve because it was obtained with synthetic "data" computed from equation (7); we call such a smooth curve a "deviation pattern." The way in which its shape reflects the inherent natures of the equations may be deduced by studying the curves in Figure 9, from which Figure 8 may easily be derived.

The shape of this curve is characteristic in the topological

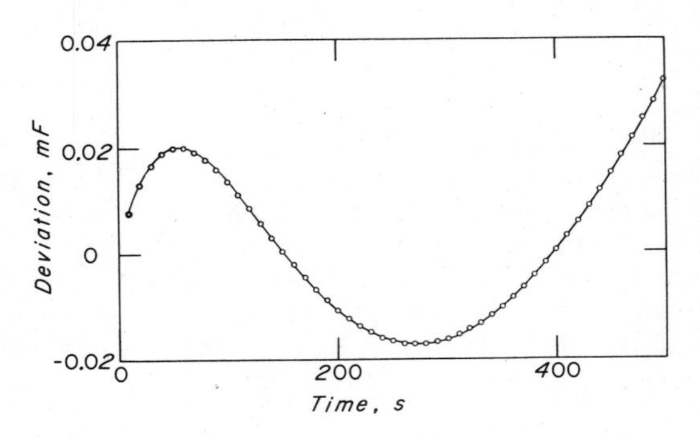

Figure 8. Deviation pattern arising from the assumption that the dependence of Q on t is described by equation (2b) when it is actually described by equation (7). The ordinate of each point is the difference between the measured value of Q and the value calculated from equation (2b), using the values of Q_∞ and s that yield the best fit to that equation.

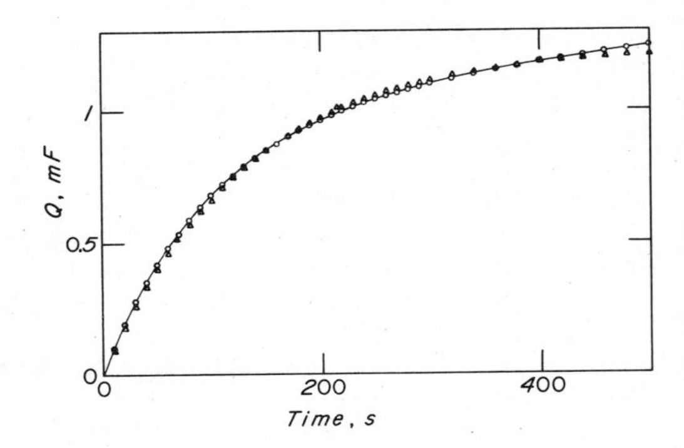

Figure 9. Rationale of the deviation pattern shown in Figure 8. The solid curve and the open circles represent the calculated dependence of Q on t according to equation (7) with $Q_{\infty, corr.} = 1$ mF, $s = 0.01$ s^{-1}, and $i_{f,c} = 5 \times 10^{-4}$ mF s^{-1}. The triangles show the best fit that can be obtained to equation (2b), which gives $Q_\infty = 1.236$ mF and $s = 7.78 \times 10^{-3}$ s^{-1}.

sense; its amplitude is altered by changing the value of $i_{f,c}$ or Q, and its period is altered by changing the value of s, but the eye and brain are so constructed that changes do not affect the recognizability of the pattern.

When random errors are also involved, as of course they always are in dealing with real data, the deviation plot consists of random errors superimposed on the pattern. If the random error is much smaller than the amplitude of the pattern, it is still easy to recognize the pattern on visual inspection, and it can even be distinguished from some generally similar one corresponding to another equation different from equation (7). With any given random error, decreasing the amplitude of the pattern (e.g., by decreasing the value of $i_{f,c}$ while holding the values of the other parameters constant) first makes it more difficult to distinguish the underlying pattern from another similar one. Decreasing the amplitude still farther makes it difficult, and eventually impossible, to detect the existence of the pattern. This is a complicated way of saying that a systematic error may be too small to detect. In our experience the difficulty of perceiving a pattern visually becomes severe if the random error is as large as about half of the peak-to-peak amplitude of the deviation pattern on which it is superimposed.

Different deviation patterns are frequently easy to distinguish visually if the random error is very small. For example, the deviation pattern in Figure 10, which is obtained on fitting data that actually correspond to the mechanism

$$\begin{aligned} A + n_1e &= P \quad \text{(mass-transfer constant} = \underline{s}_1) \\ B + n_2e &= Q \quad \text{(mass-transfer constant} = \underline{s}_2) \end{aligned} \tag{16}$$

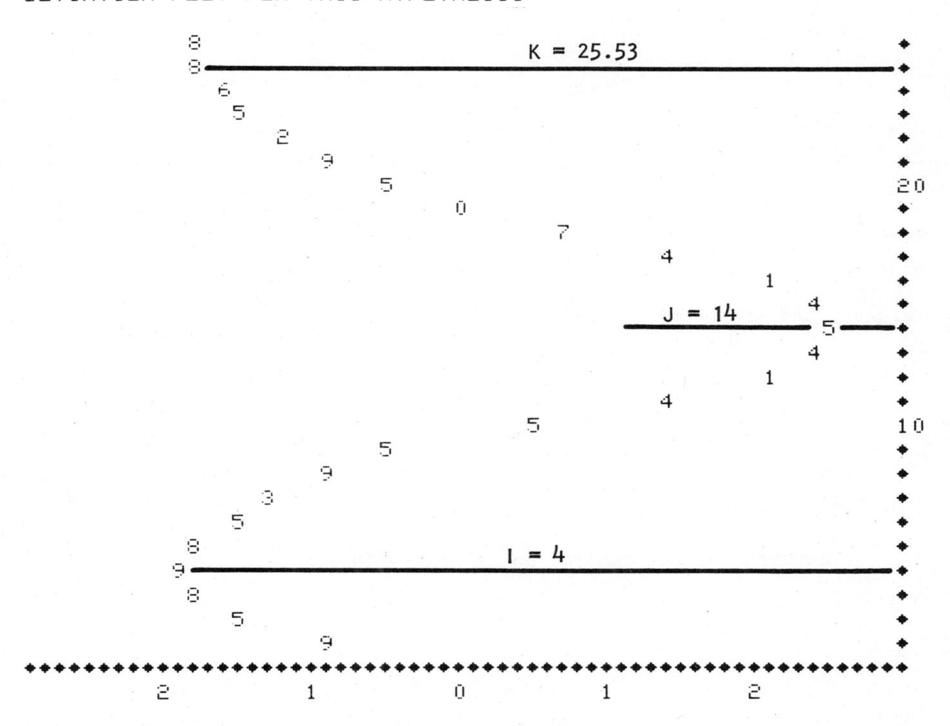

Figure 10. Deviation pattern arising from the assumption that the mechanism of a controlled-potential electrolysis is described by equation (1) when it is actually described by equations (16). As compared with the plots in Figures 8 and 9, this is rotated through 90°, with the ordinal numbers of the data points (which increase monotonically with time) being plotted along the vertical axis and deviation being plotted along the horizontal axis. In addition, the abscissa of this plot is compressed at long times because the increments of time between successive points increases as the electrolysis proceeds. Each number in the body of the figure is equal to the tenths digit in the value of the deviation, which is thus equal to 1.8 units for the last point shown. The significances of I, J, and K are explained below.

EXAMINATION OF DEVIATION PLOTS

Are there three consecutive positive deviations whose sum exceeds $4.5s_Q$?

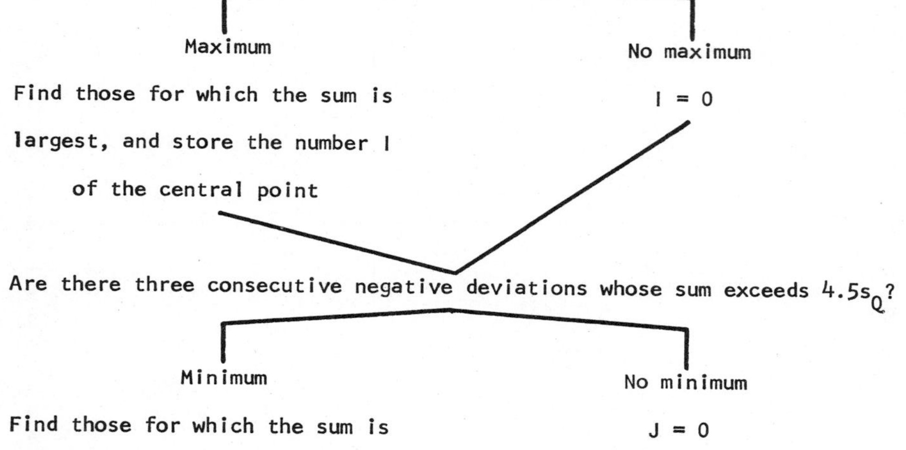

largest, and store the number J

of the central point

Excluding any feature already found, identify the three consecutive points for which the sum S is largest. Store the value and sign of S and the number N of the central point as K = (N + |S|/10)*SGN(S).

Figure 11. Scheme for identifying maxima and minima on deviation plots and storing their locations and amplitudes

(with $c^o_A = c^o_B$, $\underline{n}_1 = \underline{n}_2$, and $\underline{s}_1 = 1.2\underline{s}_2$) to equation (2$\underline{b}$), is quite different from the one in Figure 8. Near the end of the electrolysis the slope of the former approaches zero, while that of the latter increases continuous y. This is so subtle a difference, however, that it would be completely masked by even a quite small random error, and the same thing is true of many other closely similar patterns. The program therefore embodies a scheme for recognizing patterns that is sensitive only to gross differences of shape that cannot easily be hidden by random errors.

This scheme is shown diagrammatically in Figure 11. The successive values of ($\underline{Q}_{meas.} - \underline{Q}_{calc.}$) are inspected to see whether the sum of three consecutive ones exceeds $4.5\underline{s}_Q$, where $\underline{s}_Q$ is the estimated standard error of a single measured value of $\underline{Q}$. In the absence of any systematic error, the probability that any one deviation would equal or exceed $1.5\underline{s}_Q$ is equal to 0.0668; the probability that three consecutive ones would do so is equal to $(0.0668)^3 = 3 \times 10^{-4}$. If there are no three consecutive points that satisfy this criterion, it is concluded that there is no maximum on the deviation plot, and a quantity $\underline{I}$ is set equal to zero. If there are three such points, the three giving the largest sum of the values of $(\underline{Q}_{meas.} - \underline{Q}_{calc.})$ are identified, and $\underline{I}$ is set equal to the ordinal number $\underline{N}$ of the middle one of these points.

Next a minimum is sought in a similar way. For this purpose a minimum is arbitrarily defined as a set of three consecutive points for which the sum of the values of $(\underline{Q}_{meas.} - \underline{Q}_{calc.})$ is more negative than $-4.5\underline{s}_Q$. The deepest minimum is identified and a second quantity $\underline{J}$ is set equal to the ordinal number $\underline{N}$ of the point at its center; if there is no minimum $\underline{J}$ is set equal to zero.

If the quantities $\underline{I}$ and $\underline{J}$ differ from zero, their values identify the largest maximum and the largest minimum on the deviation plot. Values of $\underline{N}$ close to either $\underline{I}$ or $\underline{J}$ are excluded from a third search, which identifies the set of three consecutive points for which the absolute value of the sum $\underline{S}$ of the deviations is largest. When this has been done, a third quantity $\underline{K}$ is set equal to $[\underline{N} + \mathrm{ABS}(\underline{S}/10)] \cdot \mathrm{SGN}(\underline{S})$, where $\underline{N}$ is the ordinal number of the middle one of this set of three points. From the resulting values of $\underline{I}$, $\underline{J}$, and $\underline{K}$ it is possible to tell whether the plot has two maxima and one minimum or two minima and one maximum, where these features lie with respect to its $\underline{t}$-axis, and how large the lesser maximum or minimum is.

Two possibilities now arise. One is that $\underline{I} \cdot \underline{J}$ is equal to zero, which logically means that the plot lacks either a maximum or a minimum according to the above definitions, and usually in practice means that it lacks both. In this case the hypothesis that equation (2$\underline{b}$) accounts for the curve is accepted, and a final segment of the program is called into core to provide an appropriate message. We shall return to this in a later paragraph.

The other possiblity is that $\underline{I} \cdot \underline{J}$ has some finite value. In this case the hypothesis is rejected. Since the probability that either $\underline{I}$ or $\underline{J}$ will be assigned a finite value if the hypothesis is correct is only 3×10^{-4}, the probability that the hypothesis will be incorrectly rejected is only 10^{-7}. The smallness of this figure reflects, but considerably exaggerates, our bias in favor of accepting the simplest tenable hypothesis, and it will probably be revised upward in further work.

Normally it would be possible to choose either of two alternative hypotheses at this stage: one if the deviation plot had two maxima and one minimum, and another if it had two minima and one maximum. However, it is an unusual feature of the present problem that curves of both these shapes can arise from a continuous faradaic background current. If this has the same sign as the current for the reduction or oxidation of A, the plot has two maxima and one minimum and resembles Figure 9. If it has the opposite sign, the plot has two minima and one maximum. Continuous faradaic backgound currents are rarely positive (According to a recent IUPAC recommendation, positive currents correspond to anodic processes at the working electrode, and negative ones to cathodic processes. This recommendation is followed here, but it is directly opposite to established practice among electroanalytical chemists.), but since the process being studied may be either cathodic or anodic, provision must be made for both possibilities. Hence it is impossible to effect a branch at this point, and any finite value of the product $\underline{I} \cdot \underline{J}$ leads to the tentative adoption of the hypothesis embodied in equation (7).

A fit to this equation is begun by deriving an estimate of $\underline{i}_{f,c}$ from the coordinates of two points near the end of the curve, and applying appropriate small corrections to the values of $\underline{Q}$ and $\underline{s}$ obtained in the preceding fit. Regression onto equation (7) is then effected in the ordinary way, and a new deviation plot is constructed and examined as before. The hypothesis is accepted if $\underline{I} \cdot \underline{J} = 0$; if $\underline{I} \cdot \underline{J}$ has a finite value either of two branches may be pursued, depending on what the value of $\underline{K}$ reveals about the shape of the plot. For example, it was said earlier that the hypotheses represented by equations (7) and (16) will probably be indistinguishable on a deviation plot constructed from equation (2$\underline{b}$) unless the random errors of measurement are very small. However, the shapes of the two patterns are not actually the same, and the difference is clearly revealed on constructing a new plot from equation (7). If there is actually a continuous faradaic background current this plot will consist of points randomly scattered around the $\underline{t}$-axis. If there are actually two substances undergoing simultaneous but independent reductions, the points will be scattered around the pattern shown in Fig. 12. The sensitivity of this procedure for detecting simultaneous processes may be gauged from the fact that there was a difference of only 20% between the values of the two rate constants employed in calculating the "data" on which this Figure is based.

We turn now to the segments of the program that provide the final output. For the sake of brevity we shall discuss only the one that corresponds to acceptance of the hypothesis represented by equation (2$\underline{b}$).

If this equation yields a satisfactory fit, the mechanism may be the one described by equation (1), but there are other possibilities as well. The starting material A may undergo a pseudo-first-order reaction with the solvent or some other major constituent of the solution Z, yielding a product Q that is not electroactive:

$$A + \underline{n}\,\underline{e} = P; \quad A + Z = Q \qquad (17)$$

In this event the current will decay exponentially with time and equation (2*b*) will be obeyed, but the value of $\underline{n}_{app.}$ calculated from equation (5) will be smaller than that of $\underline{n}$. There may be an "induced" reduction of Z:

$$A + Z = AZ \; ; \; AZ + \underline{n}'\underline{e} = A + Z' \; ; \; A + \underline{n}\,\underline{e} = P \qquad (18)$$

Equation (2*b*) will again be obeyed, but the value of $\underline{n}_{app.}$ will exceed that of $\underline{n}$. The starting material may be reduced in two different ways to yield two different products:

$$A + \underline{n_1 e} = P \; ; \; A + \underline{n_2 e} = Q \qquad (19)$$

Equation (2*b*) will be obeyed, and $\underline{n}_{app.}$ may be either smaller or larger than the value expected *a priori*.

The existence of these possibilities makes it necessary to compare the value of Q_∞ with that of $\underline{n}\,\underline{N}^{o}_{A}$ [equation (3)]. The variance of Q_∞ depends on the standard error of the individual

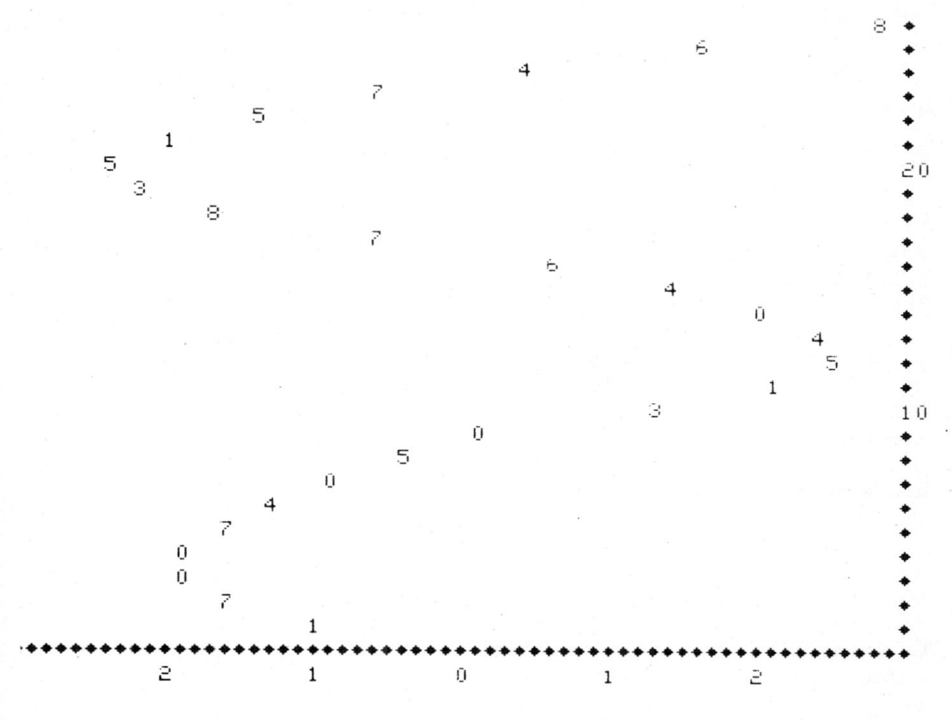

Figure 12. Deviation pattern arising from the assumption that the dependence of Q on t is described by equation (7) when the mechanism is actually described by equations (16)

```
AN ACCEPTABLE FIT IS OBTAINED ON ASSUMING THE I-T CURVE TO BE
 EXPONENTIAL.

THE DEVIATION PLOT IS FEATURELESS AND THE STANDARD ERROR OF
 THE FIT IS CONSISTENT WITH THE ESTIMATE PROVIDED.

A ONE-TAILED CHI^2 TEST INDICATES THAT MEANINGFUL NON-RANDOM
 ERRORS ARE ABSENT AT THE 98.66 % LEVEL OF CONFIDENCE.
 THIS FACT WAS NOT USED IN MAKING THE CLASSIFICATION BUT
 TENDS TO SUPPORT ITS CORRECTNESS.

CONCLUSION IS THAT THE PROCESS OCCURS IN A SINGLE STEP AND
 THAT ITS RATE IS MASS-TRANSFER-CONTROLLED:
                        A + 1 E = P
 WITH N(APP.)= .9985316 , IN ACCEPTABLE AGREEMENT WITH
 THE EXPECTED VALUE, AND S= .01000633 S^-1.

IF THIS CONCLUSION IS CORRECT, THE CLASSIFICATION AND THE
 VALUE OF N(APP.) WILL BE UNAFFECTED BY VARYING THE INITIAL
 CONCENTRATION OF A, THE STIRRING EFFICIENCY, AND E(W.E.) AS
 LONG AS THIS REMAINS ON THE PLATEAU OF THE WAVE OF A.

THERE MAY ALSO BE A PRIOR EQUILIBRIUM BETWEEN A AND ITS
 PRECURSOR Y -- Y + Z = A (FAST) ; A + 1 E = P .  IN THIS
 CASE THE CLASSIFICATION SHOULD CHANGE (TO CASE 2C, P. 698)
 IF THE CONCENTRATION OF Z IS SUFFICIENTLY DECREASED.

READY
```

Figure 13. Typical final output obtained with synthetic data conforming to the mechanism described by equation (1)

values of Q and on the number of data points and the manner in which these are distributed along the t-axis (16). We chose to adopt a fixed data-acquisition schedule, which was being used by one of us for research in this area eighteen years before the present work was begun. This enabled us to calculate the ratio s_{Q_∞}/s_Q and to use it in assessing the difference between the calculated and expected values of Q_∞. If this difference is smaller than, say, $3s_{Q_\infty}$, it may be concluded that the mechanism is adequately represented by equation (1). There may in addition be a fast prior equilibrium between the electroactive substance A and some other species Y:

$$Y + Z = A(\text{fast}) \quad ; \quad A + n\,e = P \tag{20}$$

but none of the many other possibilities yet envisioned will give rise both to an exponentially decaying current and to a value of Q_∞ coinciding with the prediction of equation (3). Should the calculated and expected values of Q_∞ disagree, the sign of their difference is used to produce a message identifying the most likely possibilities and telling how they can be tested. Typical examples of the final output from this portion of the program are shown in Figures 13 and 14. There are many other possibilities, but a full discussion of them would be impossible in the space available.

Results and Discussion

How reliable are the classifications effected by such a program? This one has been tested in two ways: with experimental data for systems believed to be thoroughly well understood, and

```
AN ACCEPTABLE FIT HAS BEEN OBTAINED BY ASSUMING THAT THE I-T
 CURVE IS THE SUM OF AN EXPONENTIALLY DECAYING CURRENT AND A
 CONSTANT ONE.

THE DEVIATION PLOT IS FEATURELESS AND THE STANDARD ERROR OF
 THE FIT IS CONSISTENT WITH THE ESTIMATE PROVIDED.
A ONE-TAILED CHI^2 TEST INDICATES THAT MEANINGFUL NON-RANDOM
 ERRORS ARE ABSENT AT THE 100 % LEVEL OF CONFIDENCE.
 THIS FACT WAS NOT USED IN MAKING THE CLASSIFICATION BUT
 TENDS TO SUPPORT ITS CORRECTNESS.

THE CONSTANT CURRENT MAY BE A CONTINUOUS FARADAIC BACKGROUND
 CURRENT EQUAL TO 9.645899E-6 MF/S = .9305274 MA.

THIS CONCLUSION SHOULD BE CONFIRMED BY OTHER EXPERIMENTS IN
 WHICH THE STIRRING EFFICIENCY, INITIAL CONCENTRATION OF A,
 AND E(W.E.) ARE VARIED.  IF IT IS CORRECT,
 THE CLASSIFICATION AND VALUE OF I(F,C) WILL BE INDEPENDENT
 OF STIRRING EFFICIENCY AND CONCENTRATION OF A, BUT I(F,C)
 WILL VARY EXPONENTIALLY WITH E(W.E.). MOREOVER, A VIRTUALLY
 EQUAL STEADY-STATE CURRENT SHOULD BE OBTAINED IN ELECTROLY-
 SES OF THE SUPPORTING ELECTROLYTE ALONE AT THE SAME E(W.E.).

THE EXPONENTIALLY DECAYING CURRENT CORRESPONDS TO A
 SINGLE-STEP MASS-TRANSFER-CONTROLLED PROCESS, A + NE = P ,
 WITH N(APP.)= .9987849 AND S= 9.995575E-3 S^-1

THIS CONCLUSION SHOULD BE CONFIRMED BY OTHER EXPERIMENTS IN
 WHICH THE STIRRING EFFICIENCY, INITIAL CONCENTRATION OF A,
 AND E(W.E.) ARE VARIED.  IF IT IS CORRECT,
 THE CLASSIFICATION AND THE VALUE OF N(APP.) WILL BE UNAF-
 FECTED BY THESE CHANGES.

THERE MAY ALSO BE A FAST PRIOR EQUILIBRIUM BETWEEN A AND ITS
 PRECURSOR Y: Y + Z = A (FAST); A + NE = P.  IN THIS CASE THE
 CLASSIFICATION WILL CHANGE (TO CASE 2C, P. 698) IF THE CON-
 CENTRATION OF Z IS SUFFICIENTLY DECREASED.


READY
```

Figure 14. Typical final output obtained withe synthetic data conforming to equation (7)

```
3. AN APPARENTLY CONSTANT CURRENT HAVING THE OPPOSITE SIGN
   FROM THE CURRENT FOR REDUCTION OR OXIDATION OF A, AND
   EQUAL TO APPROXIMATELY 3.500000E-6 MF/S = .35 MA,
   WHICH IS TOO SMALL FOR POSITIVE IDENTIFICATION BY THIS
   PROGRAM. A CONSTANT CURRENT OF THE MAGNITUDE GIVEN ABOVE SHOULD
   BE OBSERVED IN THE BLANK SOLUTION AT THE SAME E(W.E.),
   AND ITS VALUE SHOULD BE SUPPLIED IN THE INITIAL DIA-
   LOGUE WHEN THIS PROGRAM IS RE-RUN.
```

Figure 15. A portion of the final output obtained when $i_{f,c}/(i^{\circ} - i_{f,c})$ *is between* 3×10^{-4} *and* 8.5×10^{-4}*. The value given for the constant current is calculated from an empirical relation between* $i_{f,c}/(Q_{\infty}s - i_{f,c})$ *and* $Q_{\infty, calc.} - Q_{\infty, true}$.

with synthetic data obtained by superimposing normally distributed random errors on the theoretical values calculated from exact equations. As the work is still in progress, we can provide a full report only on discriminations between the simplest mechanism [equation (1)] and the case in which a continuous faradaic background current is also involved.

There are three different results that may be obtained, and it is convenient to describe them in terms of the ratio of the continuous faradaic background current to the initial current for the reduction of A. This ratio corresponds to $\underline{i}_{f,c}/(\underline{i}^{o}-\underline{i}_{f,c})$ in equation (6). The values cited in the following paragraphs reflect certain arbitrary decisions made in constructing the program, and have no absolute significance. Nevertheless they correspond surprisingly well to the levels at which an informed human chemist would draw lines between reasonable assurance and substaintial doubt in interpreting the same data.

1. $\underline{i}_{f,c}/(\underline{i}^{o}-\underline{i}_{f,c}) \geq 0.001$. Thirty-seven sets of data fulfilling this condition were analyzed; every one was classified as involving a continuous faradaic background current - or some phenomenon that even a human chemist could not distinguish from this on the basis of a single experiment.

2. $\underline{i}_{f,c}/(\underline{i}^{o}-\underline{i}_{f,c}) \leq 0.00025$. Forty-four sets of data fulfilling this condition were analyzed; every one was classified as corresponding to a simple process - or, again, something indistinguishable from this.

3. $0.0003 \leq \underline{i}_{f,c}/(\underline{i}^{o}-\underline{i}_{f,c}) \leq 0.00085$. Twenty-four sets of data fulfilling this condition were analyzed. For each one equation (2b) provided a satisfactory fit, but yielded a value of $\underline{Q}_{\infty}$ that was not in acceptable agreement with the expected one. The hypothesis represented by equation (1) is then accepted and the final printout includes the paragraph shown in Figure 15 as one of several possible explanations of the error in $\underline{Q}_{\infty}$.

It is not surprising that the boundary between regions 2 and 3 should be as sharp as it is. There is a very direct relation between the value of $\underline{i}_{f,c}/(\underline{i}^{o}-\underline{i}_{f,c})$ and the value of $\underline{Q}_{\infty}$ obtained from a fit to equation (2b), and the standard error of $\underline{Q}_{\infty}$ is so small that even a small variation of $\underline{i}_{f,c}/(\underline{i}^{o}-\underline{i}_{f,c})$ produces a comparatively large variation of $\underline{Q}_{\infty}$.

However, we were not prepared to find that the boundary between regions 1 and 3 is equally sharp. The boundary between regions 2 and 3 is governed by purely statistical considerations, but the one between regions 1 and 3 depends on the possibility of identifying the features of a deviation pattern in the face of random errors that occasionally cloak them against visual detection. We see no way of avoiding the conclusion that deviation-pattern recognition in the fashion outlined here is no less reliable and useful than the familiar tests prescribed by classical statistics.

Conclusion

There appear to be two, and only two, reactions that can be manifested by the chemist watching the execution, or inspecting the final output, of a properly constructed classification program. Rather to our surprise, we have detected both in ourselves. One, which was mentioned above, is disbelief that the result of a human being's devoted application of understanding and intuition can be so accurately duplicated by a machine. As such programs become more common in the years to come - no matter whether they involve pattern recognition, the sort of numerical analysis outlined here, or some other technique not yet devised - this incredulity will probably be the chief deterrent to their wide and ready adoption. Those who write such programs, and who study the principles on which such programs can be based, should realize that they appear to much of the scientific community to be challenging the worth of the human being in science, and would do well to face the psychological and philosophical problems that are involved.

The other common reaction is to personify the computer executing the program. This is especially apt to happen if the numerical computations and other processes that produce and guide the final classification are concealed during execution, and if the final output includes possibilities that the watcher recognizes as being correct but did not think of or remember until the computer terminal printed them out. It is oddly easy to forget the role that the programmer has played in paving the path from initial input to final printout.

Whatever the extent to which these problems may hinder and delay the achievement of general popularity and widespread use by machine-classification procedures, there is a very real foundation for the enthusiasm that their proponents have evinced at this symposium. Believing that the length of his scientific career exceeds that of any other member of this group, the senior author of this paper wishes to express his conviction that he has seen no other development in scientific thought, knowledge, or methodology - with the single exception of on-line data acquisition and processing - that has had anything like the same promise of changing the structure of science and the professional lives of those who pursue its ends. For this reason he hopes that this symposium will swell the ranks of those who are active in its field.

Literature Cited

1. Brand, M.J.D., and G. A. Rechnitz, Anal. Chem., (1970), 42, 1170.
2. Isbell, A. F., JR., R. L. Pecsok, R. H. Davies, and J. H. Purnell, Anal. Chem., (1973), 45, 2363.
3. Ingman, F., A. Johansson, S. Johansson, and R. Karlson, Anal. Chim. Acta, (1973), 64, 113.
4. Barry, D. M., and L. Meites, Anal. Chim. Acta, (1974), 68, 435.
5. Barry, D. M., B. H. Campbell, and L. Meites, Anal. Chim. Acta. (1974), 69, 143.
6. Campbell, B. H., and L. Meites, Talanta, (1974), 21, 393.
7. Meites, L., and E. Matijević, Anal. Chim. Acta, (1975), 76, 423.
8. Meites, L., Pure Appl. Chem., (1969), 18, 35.
9. Meites, L., "Controlled-Potential Electrolysis," Chap. IX in "Physical Methods of Chemistry (Vol. I of 'Techniques of Chemistry,' ed. by A. Weissberger and B. W. Rossiter) Part IIA. Electrochemical Methods," Wiley-Interscience, New York, 1971.
10. Bard, A. J., and K. S. V. Santhanam, "Application of Controlled Potential Coulometry to the Study of Electrode Reactions," in "Electroanalytical Chemistry," ed. by A. J. Bard, Marcel Dekker, Inc., New York, Vol. 4, 1972.
11. Lingane, J. J., "Electroanalytical Chemistry," Interscience, New York, 2nd ed., 1958, pp. 222-9.
12. Meites, L., "The General Multiparametric Curve-Fitting Program CFT4," Computing Laboratory of the Department of Chemistry, Clarkson College of Technology, Potsdam, N.Y., 1976.
13. Meites. L., and L. Lampugnani, Anal. Chem., (1973), 45, 1317.
14. Meites, L., and Barry, D. M., Talanta, (1973), 20, 1173.
15. Campbell, B. H., L. Meites, and P. W. Carr, Anal. Chem., (1974), 46, 386.
16. Meites, L., Anal. Chim. Acta, (1975), 74, 177.

8

Examples of the Application of Nonlinear Regression Analysis to Chemical Data

Y. C. MARTIN and J. J. HACKBARTH

Abbott Laboratories, North Chicago, IL 60064

Nonlinear regression analysis is a powerful mathematical tool which has been used by a few chemists (1-8), but which has not achieved widespread application in chemistry. It is the purpose of this communication to illustrate some of the circumstances in which we have found the method to be useful. The objective is to encourage others to use this technique for other data-fitting problems.

Introduction to Nonlinear Regression Analysis

What is nonlinear regression? How does it differ from linear regression? Linear regression analysis is the process of finding the least-squares best fit of a set of data to a uni- or multidimensional equation in which the parameters (coefficients) to be fit are linear functions of the observed properties. The simplest linear regression analysis involves fitting data to a single parameter; as the name implies, the equation is that of a straight line:

$$Y_i = b_0 + b_1X_i + \varepsilon \qquad \text{eq. 1}$$

In equation 1, Y_i is the observed value of the dependent variable of observation i, X_i is the value of the independent variable for observation i, b_0 is the intercept of the line on the Y axis, b_1 is the slope of the line, and ε is the error.

Nonlinear regression analysis is the process of finding the least-squares best fit of a set of data to an equation which is not linear in the parameters to be fit. A very simple example is:

$$\log(Y_i) = \log(1 + aX_i^{b}) + \varepsilon \qquad \text{eq. 2}$$

Equation 2 is nonlinear in a and b, the parameters to be fit.

Most chemical relationships are nonlinear: one familiar example is the fraction of an acid ionized as a function of pH and pK_a:

$$\alpha = \left[\frac{1}{10^{pK_a - pH} + 1} \right] \qquad \text{eq. 3}$$

If some physical property were linearly related to α, then the observed variables would be the physical property in question and pH. The pK_a would be the parameter to be fit.

How does one determine what is the "best" fit in the nonlinear case? As with linear regression, the least-squares criterion for best fit is commonly used. It is defined as that choice of value for the adjustable parameters (b_0 and b_1 in Equation 1 or a and b in Equation 2) such that the sum of squared difference between the observed Y_i's and those calculated on the basis of the X_i's is a minimum. Mathematically, this involves taking the partial derivative with respect to each of the parameters of the equation for the sum of the squared differences, setting this derivative equal to zero, and solving for the parameter. In the linear case this all works very well--the values of b_0 and b_1 can be explicitly determined. Nonlinear equations do not yield such an easy solution for the minimum sum of squares: hence, in order to find the best fit to a nonlinear equation an iterative procedure must be used. Hence one starts with a set of best guesses for the values of the parameters to be fit. The sum of squared deviations between observed and calculated Y_i's is then calculated. By some algorithm another set of (better) estimates is chosen and the sum of squared deviations is calculated from these values. This process continues until, by some pre-established criterion, further changes in the estimates do not decrease the sum of squared deviations from the fit.

From this brief description it can be seen that nonlinear regression analysis suffers from several apparent disadvantages compared to linear regression. An initial estimate of the parameter values must be supplied, an algorithm for finding the minimum sum of squares must be provided, and many calculations of this sum of squares are required for the solution to one problem. In view of these difficulties, the traditional method for fitting a nonlinear equation has been to transform the equation into a linear form and fit the data to this transformed equation. The disadvantages of such a linearization strategy are that it may involve hours of algebraic manipulation, that frequently assumptions must be made as to the range of the data or the importance of terms in a sum, and that the resulting equation implicitly weights the data in a manner which may not be consistent with experiment.

On the other hand, advances in computer technology have made these disadvantages of nonlinear regression analysis relatively unimportant. Initial estimates are easily determined with standard graphics techniques in which the data and calculated curve are displayed on a cathode ray tube. Algorithms to find minima and maxima are easy to implement. Finally, the computations of many sums of squares is a trivial task for a computer. The major advantages of nonlinear regression analysis are that one fits data to the equation as derived and that simplifying assumptions are not necessary.

Computer Programs

As noted above, in the analysis of nonlinear problems we have found it convenient to be able to display the data and a calculated curve on a cathode ray tube. The parameters of the calculated curve may then be altered until visually there is a reasonable fit to the data. Hence, initial searching of the parameter space for estimates is not difficult, in fact it often reveals unsuspected facets of the data. A companion program takes the same data file and generates instructions for a Calcomp plotter to make a hard copy of what was seen on the screen. We have written programs for two dimensions (one independent variable) and three dimensions (two independent variables).

We use a simplex method to find the minimum sum of squares (9,10). The algorithm includes expansion and contraction of dimensions of the simplex. The statistical properties of the best fit are calculated by the equations used in the BMD P-series (11).

General Description of Examples

The following discussion illustrates the two principle types of application which we have made of nonlinear regression analysis. The first type of application is the calculation of physical properties of a molecule from experimental observations. We have chosen as an example the calculation of the pK_a's of a dibasic substance from measurements of absorbance *vs* pH. The other two examples are from our work on the quantitative relationship between physical properties and biological potency of drug analogs. Variation in antibacterial potency of nitrophenols and erythromycins as a function of structure and pH of the test are examined.

Example 1: pK_a's of a Dibasic Substance from Absorbance Measurements

Recently we wanted to know the pK_a's of the following compound:

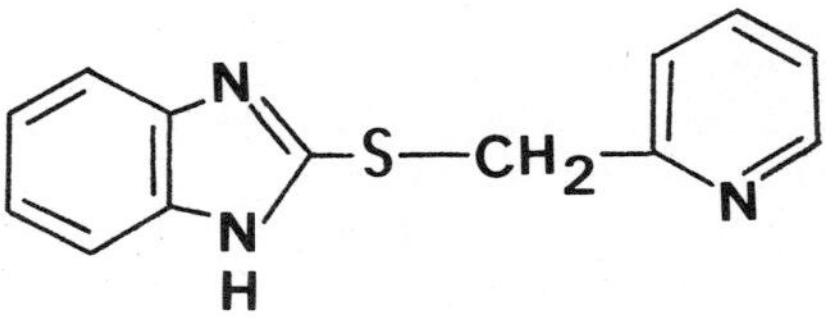

Since it is not soluble enough to titrate, we measured its ultraviolet spectrum with changes in degree of protonation, _ie_ as a function of pH. Essentially the method described by Albert and Serjeant were followed (12). That is, the ultraviolet spectrum of the compound was recorded in buffer solution at several pH's near the suspected pK_a's. Figure 1 shows the absorbance at 304 nm as a function of pH. Because measurable changes occurred over such a wide pH interval, we concluded that the absorbance change reflected more than one pK_a.

The first nonlinear function we attempted to fit was that given by Albert and Serjeant (12):

$$A = \left[\frac{c_t a_d [H^+]^2 + c_t a_m [H^+] K_1 + c_t a_n K_1 K_2}{[H^+]^2 + [H^+] K_1 + K_1 K_2} \right] \qquad \text{eq. 4}$$

in which c_t is the total concentration of the compound; a_d, a_m, and a_n are the molar absorbtivities of the dication, monocation, and neutral forms respectively; and K_1 and K_2 are the first and second acid dissociation constants. This form of the equation led to problems in fitting, particularly with data from tribasic analogs. Solution of Equation 4 in terms of $[H^+]$ led to an algebraically very complex relationship. So we manipulated Equation 3 further until we realized the following:

$$A([H^+]^2 + [H^+]K_1 + K_1K_2) = c_t a_d [H^+]^2 + c_t a_m [H^+] K_1 + c_t a_n K_1 K_2$$

or

$$0 = (c_t a_d - A)[H^+]^2 + (c_t a_m - A)[H^+]K_1 + (c_t a_n - A)K_1K_2$$

Since all terms add up to a sum of zero, the sum of those with a positive sign must equal the sum of those with a negative sign. The former sum is labelled "POS", and the latter, "NEG". Hence:

$$POS = -NEG$$

or

$$1 = \frac{POS}{-NEG}$$

or

$$0 = \log \frac{POS}{-NEG}$$

Hence we chose to fit the following function:

$$pH_{calc} = \log \left[\frac{POS}{-NEG} \right] + pH_{obs} \qquad \text{eq. 5}$$

The computer program tests the sign of each term in the equation for each data point, sums the positive and negative terms separately, and places the sum of the positive values in the numerator and the sum of the negative values in the denominator.

From this type of analysis of the data, the following values were calculated:

$$pK_1 = 2.31 \pm .02$$
$$pK_2 = 4.45 \pm .02$$
$$c_t a_d = 0.508 \pm .001$$
$$c_t a_m = 0.316 \pm .002$$

$c_t a_n$ was experimentally established from measurements at high pH to be 0.181. The statistics of fit are:

$R^2 = .9997$, $s = .0253$, with 6 degrees of freedom.

Figure 1 shows the curve calculated on the basis of this fit. From every standpoint, the use of nonlinear regression analysis allowed the maximum information to be gained from the data. The precision of fit is satisfactory considering the low number of experimental points involved and the closeness of the two pK_a's.

We have used this method of calculation to fit the pK_a's of tribasic substances from absorbance measurements at several pH's, and also to fit the pK_a's of the individual

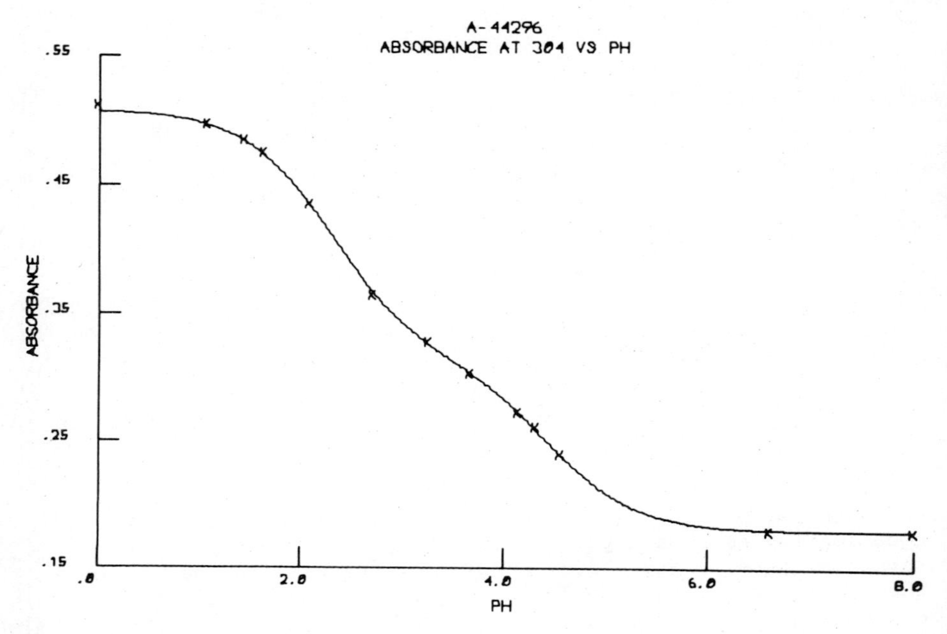

Figure 1. The variation in absorbance of Compound I at 304 nm as a function of pH. The curve is calculated from Eq. 4, $pK_1 = 2.31$, $pK_2 = 4.45$, and the absorptivity of the diprotonated species times the concentration equal to 0.580, that of the monoprotonated species equal to 0.316, and that of the nonprotonated species equal to 0.181.

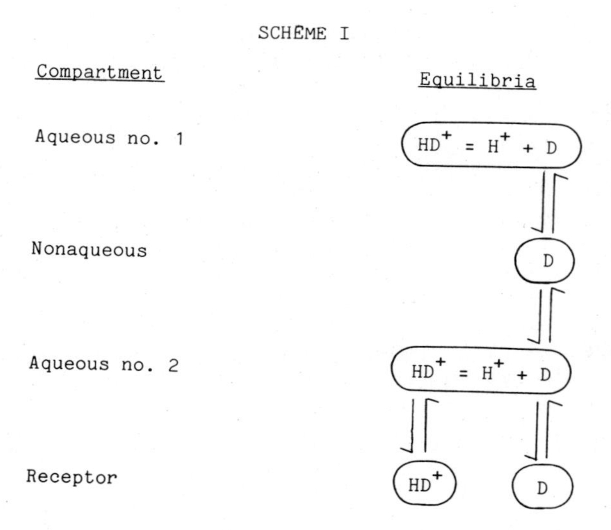

amines of polybasic molecules from the pH dependence of the change in the carbon-13 NMR chemical shift.

Example 2: Antibacterial Potency of Nitrophenols

Our principle research is in the analysis of the relationships between chemical and biological properties of compounds. We have recently become aware of the importance of the form of the equation to which the data are fit. Specifically, a consideration of the general properties of the biological system in which the data was generated coupled with a statement of the linear free energy assumptions which may be made about the structure-activity relationships of each part of this biological system can lead to some very useful insights into the form of the equation which should be used to analyze the data.

Our first publication on such model-based equations concerned biological systems in which the drug is equilibrated between several compartments; this model applies principally to *in vitro* tests (13). The equation derived for a four compartment model (Scheme I) is sufficient to correlate the data from the examples discussed in this paper. Compartment one is the aqueous compartment outside of the bacteria, that is, the medium; compartment two is the aqueous compartment inside of the bacteria; compartment three is a nonaqueous, nonreceptor compartment; and compartment four is the receptor. Compartments one and two become identical if their pH's are identical. From simple linear free energy assumptions about the relationship between binding in a compartment and hydrophobicity of the various analogs as measured by the octanol-water partition coefficient (*P*), the following equation may be derived:

$$\log(1/C) = \log\left[\frac{1 + Z\dfrac{\alpha_2}{1-\alpha_2}}{1 + Z\dfrac{\alpha_2}{1-\alpha_2} + dP^c + \dfrac{1}{P^b}\left[\dfrac{a_1}{1-\alpha_1} + \dfrac{a_2}{1-\alpha_2}\right]}\right] + X \qquad \text{eq. 6}$$

The symbol α indicates the fraction of the compound in the ionized form at the pH of the particular compartment. The dependent variable is the negative logarithm of the concentration, C, of the compound required to produce the defined biological response. The independent variables are α_1 and *P*. Finally, the parameters to be fit in the regression analysis are Z, a, b, c, and d. From the model it may be shown that Z is the relative affinity for the receptor of the ion

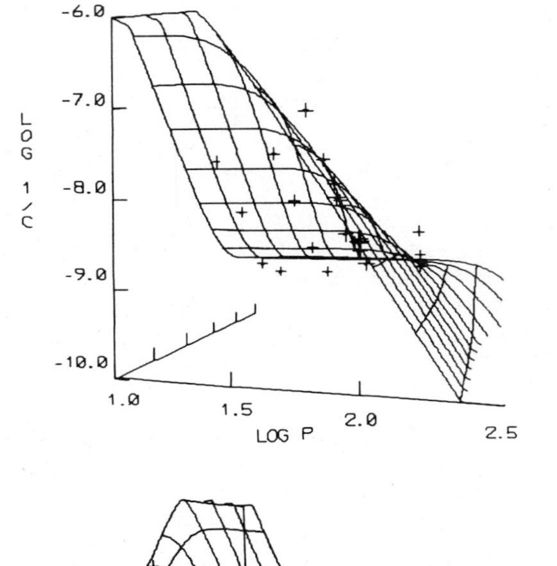

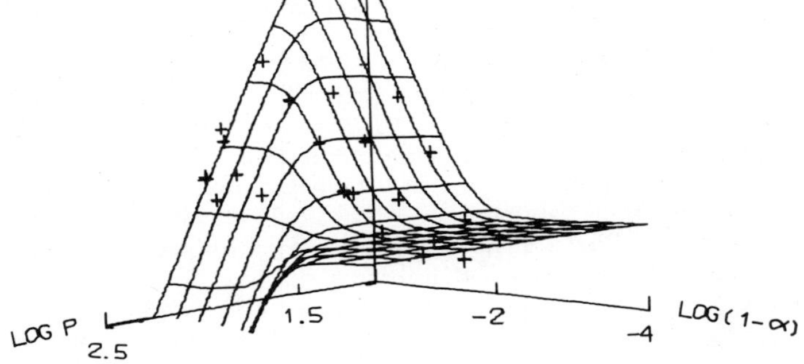

Figure 2. Two views of the variation in antibacterial activity of nitrophenols as a function of logP *and* $log(1 - \alpha)$*. The curve is calculated from Eq. 7.*

compared with that of the neutral form of a drug. A, b, c, and d are related to the extrathermodynamic relationships between binding constants to the nonaqueous and receptor compartments and log<u>P</u>.

A series of six nitrophenols had been tested <u>vs E. coli</u> at pH's 5.5, 6.5, 7.5, and 8.5 (14). All 24 data points fit the following equation (15):

$$\log(1/C) = \log\left[\frac{1 + Z\dfrac{\alpha}{1-\alpha}}{1 + Z\dfrac{\alpha}{1-\alpha} + dP^{c} + \dfrac{1}{1-\alpha}}\right] + X \qquad \text{eq. 7}$$

in which:

$$\log(Z) = -3.60 \pm 0.17$$
$$c = 4.95 \pm 0.81$$
$$\log(d) = -6.96 \pm 1.61$$
$$X_* = 5.41 \pm 0.10$$
$$pK_a = 3.26 \pm 0.73$$

The statistics of fit are:

$R^2 + 0.909$, $s = 0.209$, and $n = 24$.

Figure 2 is a plot of log(1/C) as a function of log<u>P</u> and log(1-α).

These parameters correspond to a case in which there is no variation in hydrophobic bonding to the receptor within the series. However, there is a variation in hydrophobic bonding to the nonaqueous compartment. Only one aqueous compartment (at the pH of the external medium) is needed to fit the data. A slightly better fit is found when the pK_a of picric acid is allowed to vary; this pK_a is indicated with an asterisk above. The value of Z indicates that the ionic form of any nitrophenol has 4000 X less potency than the neutral form of the same compound; this value is significantly different from zero.

<u>Example 3: Antibacterial Potency of N-benzyl Erythromycins (15)</u>

In this series one of the hydrogen atoms of the dimethylamino group of erythromycin, structure below, was replaced with a substituted phenyl group.

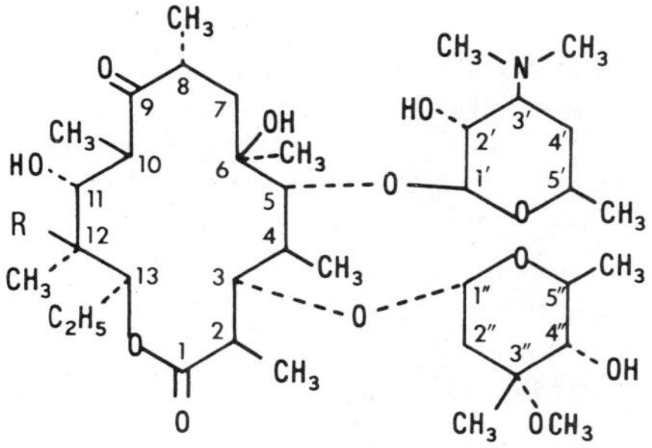

IA R=OH Erythromycin A

IB R=H Erythromycin B

The log relative potencies of these compounds as measured at pH 6.00, 7.00, and 7.65 is well fit by the following equation:

eq. 8

$$\log(1/C) = \log\left[\frac{1 + 10^{pK_a-6.0}}{1 + 10^{pK_a-6.0} + a(1 + 10^{pK_a-pH_1})}\right] + eEs + X$$

in which:

$\log(a) = 1.30 \pm 0.23$
$X = 2.64 \pm 0.17$
$e = 0.303 \pm 0.065$

The statistics of fit are:

$R^2 = .844$, $s = 0.186$, $n = 38$.

The substituent constant Es is the Taft steric parameter for the relative size of the substituent at position 4 of the phenyl ring. Figure 3 is a plot of potency as a function of the pK_a of the compound and the pH of the test.

Both the numerator and denominator of Equation 6 are dominated by a single term because the amount of ionic form of the drug bound to the receptor is larger than the amount of neutral form of the drug bound to the receptor, and because most of the drug added to the system remains in the external aqueous compartment. As a consequence of this it was not possible to independently fit a_1, Z, and α_2. Therefore, pH_2 was arbitrarily set at 6.0, and a hybrid constant, a, was fit:

$$a = \left[\frac{a_1 10^{pH_2}}{Z}\right]$$

Within the limits stated previously, the values of the constants found by the nonlinear regression analysis lead to certain tentative conclusions with respect to the determinants of potency in this series. First, two aqueous compartments of different pH must be considered in the analysis. Additionally, the pH of the aqueous phase within the bacteria remains relatively constant when the pH of the external phase is varied from 6.0 to 7.65. Second, the ionic form of these compounds contributes significantly to potency. It is not possible to establish the relative potency of the ion *vs* that of the neutral form, however, because of the problem discussed above. Third, there is no evidence for hydrophobic bonding of the phenyl ring and its substituents to the receptor or to an inert phase. Fourth, substituents at the *para* position decrease potency by a steric effect.

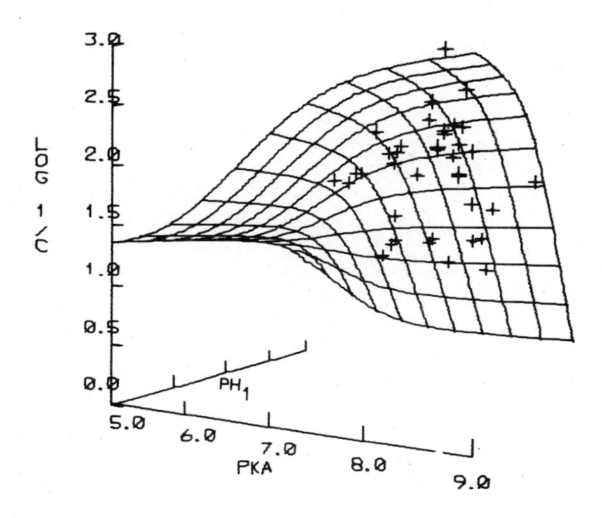

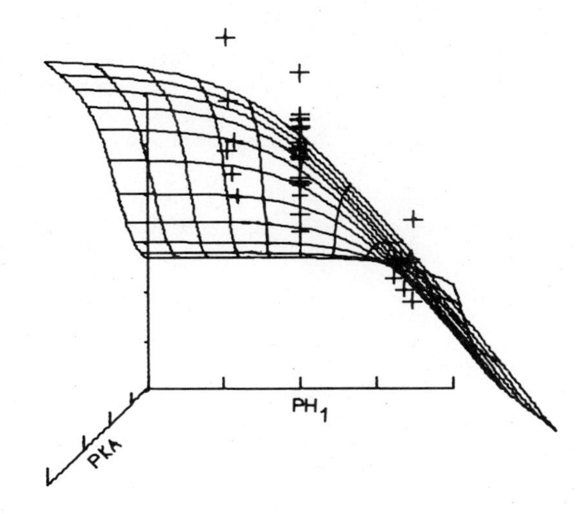

Figure 3. Two views of the variation in antibacterial activity of N*-benzyl erythromycins as a function of pH of the medium and* pK_a *of the compound. The curve is calculated from Eq. 8.*

These examples show some of the uses which we have made of nonlinear regression analysis. Once one has a little experience with a program it can easily become a routinely used and extremely helpful tool for the mathematical analysis of chemical data.

Literature Cited

1. Wentworth, W. E., J. Chem. Educ., 42 (1965) 97.
2. Jensen, R. E., R. G. Garvey, and B. A. Paulson, J. Chem. Educ., 47 (1970) 147.
3. Dye, J. L., and V. A. Nicely, J. Chem. Educ., 48 (1971) 443.
4. Barry, D. M., and L. Meites, Anal. Chim. Acta, 68 (1974) 435.
5. Barry, D. M., L. Meites, and B. H. Campbell, Anal. Chim. Acta, 69 (1974) 143.
6. Meites, L., J. E. Steuhr, and T. N. Briggs, Anal. Chem., 47 (1975) 1485.
7. Gorenstein, D. G., A. M. Wyrwicz, and J. Bode, J. Amer. Chem. Soc., 98 (1976) 2308.
8. Asleson, G. L., and C. W. Frank, J. Amer. Chem. Soc., 98 (1976) 4745.
9. Nelder, J. A., and J. Mead, Computer J., 7 (1965) 308.
10. Olsson, D. M., and L. S. Nelson, Technometrics, 17 (1975) 45.
11. Dixon, W. J., ed., "BMDP, Biomedical Computer Programs," pp 556, University of California Press, Berkeley, CA, 1975.
12. Albert, A., and E. P. Serjeant, "The Determination of Ionization Constants", pp 44-60, Chapman and Hall, London, 1971.
13. Martin, Y. C., and J. J. Hackbarth, J. Med. Chem., 19 (1976) 1033.
14. Cowles, P., and I. M. Klotz, J. Bacteriol, 56 (1948) 277.
15. Martin, Y. C., J. J. Hackbarth, and L. A. Freiberg, submitted for publication.

9

A Computer System for Structure-Activity Studies Using Chemical Structure Information Handling and Pattern Recognition Techniques

A. J. STUPER, W. E. BRUGGER, and P. C. JURS

Department of Chemistry, The Pennsylvania State University, University Park, PA 16802

The study of relationships between chemical structures and their biological activity is currently receiving a great deal of attention. The term biological activity covers a range from pharmaceuticals and drugs to agricultural chemicals such as pesticides and herbicides to toxic reactions such as those of poisions, carcinogens, teratogens, and mutagens. A variety of methods have been exploited for structure-activity studies:

(1) The semiempirical linear free energy related (LFER) or extrathermodynamic model developed by Hansch and co-workers. The LFER method is applied to homologous series of compounds that are related in that they are formed by placing substituents on a parent compound. The method depends on defining quantitative correlations between physicochemical parameters of a compound and the biological response observed. An equation of the form

$$\log (1/C) = a\pi^2 + b\pi + \rho\sigma + cE_s + d$$

is fit to the set of data using linear regression. The variables are as follows: C is the concentration of the compound necessary to produce a standard biological response; π is the difference between the logarithm of the 1-octanol/water partition coefficient of the parent compound and the substituted compound; σ is the Hammett substituent constant that provides a measure of the electronic effect on the reaction rate; and E_s is a steric factor which compares sizes of substituents to that of methyl taken as a standard.

(2) The de novo or additivity model proposed by Free and Wilson. In this approach the contributions to the parameter defining biological response by each substituent group is assumed to be additive. The equation is

$$A_i = \mu + \Sigma_j a_{j,p}$$

where μ is the overall average activity (the contribution of the constant part of the molecule, the parent structure), $a_{j,p}$ is the contribution to the activity from the jth substituent in the pth position in the parent structure, and A_i is the standard biological response for drug compound i. Regression analysis is used to obtain numerical values for the substituent contributions.

(3) Quantum mechanical methods. These methods have been used to calculate parameters to be correlated with activity and for the determination of preferred conformations of biologically active molecules.

The purpose of the present project was to apply the ADAPT computer system to specific structure-activity problems. The ADAPT computer system combines techniques of chemical structure information handling and pattern recognition for the study of chemical structure-biological activity relations. This system can be used to enter and store a set of diverse chemical structures, generate structural descriptors, and analyze them using pattern recognition methods. These three steps are illustrated in Figure 1.

Several premises are involved in this approach to the study of structure-activity relations:

- Structure and biological activity are related.
- Structures of compounds can be adequately represented as a set of molecular descriptors.
- A relation can be discovered between the structure and activity by applying pattern recognition methods to a set of tested compounds.
- The relation can be extrapolated to untested compounds.

Introduction to Pattern Recognition

Chemical and biological data are being produced at a prodigious rate. This had led to burgeoning interest in computer assisted methods for the accumulation, handling, and interpretation of these data. Standard approaches to the interpretation problem include statistical interpretation, curve fitting and model fitting. The development or verification of mathematical expressions relating independent variables and observable dependent variables is the goal of such studies. The intent is to create a model whose parameters represent quantities with physical significance. Then best values for the parameters are developed from the data by model fitting. In the absence of a mathematical model, curve fitting using general functions, *e.g.*, polynomials, can be employed. Not all problems faced by the chemist, however, lend themselves to such exacting solution: frequently, equations describing processes of interest are difficult or impossible to obtain, and a host of problems have not yielded to a satisfactory or usable theoretical explanation. In the absence of theoretically-based solutions, empirically-derived methods will often suffice to yield useful and practical solutions to complex problems.

Standard approaches to the extraction of information from complex data forms have included linear optimization, information

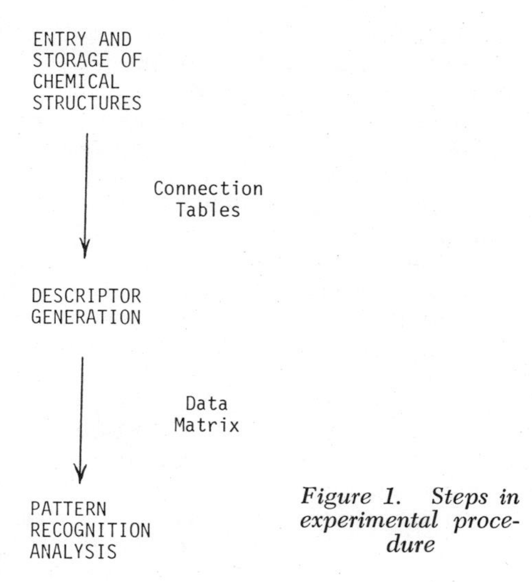

Figure 1. Steps in experimental procedure

theory, and a plethora of statistical analysis techniques. Since the early 1950's pattern recognition methods have also been applied to a variety of data interpretation problems and have paralleled the computer's growth in speed and sophistication with a corresponding expansion in scope and capacity. Pattern recognition techniques have found application in such varied fields as computer and information science, engineering, statistics, biology, physics, medicine, and physiology. Each of these disciplines has adapted the basic methods of pattern recognition to its own specific requirements.

Pattern recogniton comprises the detection, perception, and recognition of invariant properties among sets of measurements of objects or events. The purpose of pattern recognition is generally to categorize a sample of observed data as a member of the class to which it belongs. This general approach has been applied to problems from a number of diverse fields. Several excellent reviews of the pattern recognition literature have appeared which dramatize the enormous breadth of pattern recognition applications (1-5). There is a growing literature addressed to the applications of pattern recognition to chemical data interpretation.

Pattern recognition methods are uniquely suited to a variety of studies because of several novel attributes. No mathematical model is used, but rather relationships are sought which provide definitions of similarity between diverse groups of data. Pattern recognition techniques are able to deal with high dimensional data (data for which more than three measurements are used to represent each object). Such high dimensional data can not be directly visualized or displayed. In addition pattern recognition techniques can deal with multisource data or data in which the relationships are discontinuous. In multisource data each measurement can be the result of an independent generating algorithm or experiment, and each can have a different scale, origin, distribution, etc. from all the other measurements. Therefore, there will be no direct functional relationship between the measurements in multisource data as there must be, for example, in an absorbance vs. concentration plot. In applications of pattern recognition to structure-activity relations, the data is always multisource data. For problems providing multisource data, it is difficult to know in advance whether an appropriate set of measurements has been generated to effect a satisfactory solution. The generation of sufficiently informative multisource measurements can become in itself a major part of the overall pattern recognition experiment. When a number of measurements are available, pattern recognition can be used to judge their relative quality or utility with regard to specific questions. It is this ability to define relations through use of a diverse set of measurements which affords pattern recognition techniques their utility in such a wide variety of fields.

When properly used, pattern recognition techniques allow the chemist to develop criteria which relate the presence of properties to a particular sub-set of the total number of measurements. Once the important measurements are identified, they can be used to guide the development of subsequent experiments. For example, if a chemist were to find that ten structural parameters were important indicators of a particular biological effect, then he might hypothesize several as yet unstudied structures, and use the results from the pattern recognition analysis to make an educated guess as to their effects. Alternatively, the fact that the particular ten parameters were shown to be important may lead to added insights into the problem. This ability to pick a subset of the original measurements which contains the bulk of the total information content is extremely desirable. As relations between several variables are not easily deduced through observation, this is an extremely useful capability of pattern recognition.

Basic Pattern Recognition System. A general pattern recognition system for structure-activity studies must be capable of accepting numerical descriptors from the descriptor development routines performing prior feature selection preprocessing the data, and classifying the compound. A schematic representation of this basic system is shown in Figure 2. It consists of four interrelated subunits: prior feature selection, preprocessing, classification, and feedback feature selection. The prior feature selection routine accepts the data to be classified and transforms them to make the classification task easier. Then, the preprocessor attempts to pursue the following two goals simultaneously: (a) to reduce or eliminate the fraction of information contained in the raw data that is irrelevant or even confusing; and (b) to preserve sufficient information to allow discrimination among the pattern classes. The classifier operates on the transformed pattern vector to produce a classification decision. The feedback loop indicates that the pattern recognition system may use the results of its classification to develop a superior feature extraction approach. The entire pattern recognition system is generally implemented with computer software.

Classifiers. Methods of classification fall naturally into two categories: parametric and nonparametric methods. Parametric training methods consist of estimating the statistical parameters of the samples forming the training set and then using these statistical parameters for the specification of the discriminant function. Nonparametric discriminant functions are developed directly from a sample of data themselves.

Learning Machines. Data to be used in pattern recognition studies are represented as vectors, $X = (x_1, x_2, \ldots, x_n)$, where x_j represents one observation. Structures of molecules can be coded in this format using numerical descriptors for the x_j entries. For example, entries could include the molecular weight, numbers of oxygen atoms, length, volume, lipophilicity, dipole moment, number of times a particular substructure is imbedded in the structure, etc. For computational convenience an extra descriptor, whose value is set equal to a constant, is added to each pattern vector.

Data represented as vectors can be thought of either as points in an n-dimensional Euclidean space or as vectors pointing from the origin to those points, hence pattern vectors. Thus, a set of data such as a collection of mass spectra or a set of suitably encoded chemical structures can be represented as a set of n-dimensional pattern vectors. Experience shows that points representing patterns with common characteristics cluster in limited regions of the pattern space. For example, a set of points representing the molecular structures of compounds active as tranquilizers may cluster in a different region.

There is an important relationship connecting the number of points in a data set, m, and the number of descriptors per point, n, the dimensionality of the space. As shown by Nilsson (6) and by Tou and Gonzalez (7) the ability of a binary pattern classifier to separate points is high, even for random points, if m is less than twice as large as n. The probability of finding a linear decision surface capable of separating any randomly placed 50 points in a 25-dimensional space is nearly unity. Direct tests in our laboratory are in agreement with the theory of BPC's and show that one has not eliminated the possibility of meaningless training until m is two point five or three times as large as n. Thus, if one finds a separating linear decision surface for 75 points in a 25-space, then the probability is overwhelming that the separation is meaningful, and it is not a mathematical artifact.

If the clusters are dense and are far apart from each other, and if the dimensionality of the space is sufficiently low, then display or mapping techniques can be used. This is done by performing a one-to-one mapping of pattern points from the original n-dimensional space to a 2- or 3-dimensional space with as little distortion as possible. If these techniques can be successfully employed, then one can observe the clusters directly on a 2- or 3-dimensional plot.

An alternative way to investigate the structure of the set of points is to separate the clusters from one another by decision surfaces. The simplest decision surface is a hyperplane. Two clusters of points which can be completely separated by a hyperplane are said to be linearly separable. Any hyperplane has associated with it a normal vector, called here the weight vector. The weight vector consists of an ordered sequence of components, $W = (w_1, w_2, \ldots, w_n)$, which stands in one to one correspondence with the components of the patterns to be classified. Specification of the components of the weight vector is completely equivalent to specification of the position of a hyperplane decision surface.

Any pattern point in a hyperspace can be classified with respect to a hyperplane decision surface by taking the dot product between that pattern vector and the normal vector, or weight vector:

$$s = W.X = w_1x_1 + w_2x_2 + \ldots + w_nx_n = |W| \ |X| \cos \theta$$

in which θ is the angle between the two vectors. Since $|W|$ and $|X|$ are always positive, then the value of θ determines the sign of the dot product. For patterns on one side of the plane the dot product is always positive, and for patterns on the opposite side the dot product is always negative. The dot product is normally computed from the summation of pairwise products of the components of the two vectors for convenience. The correspondence between category 1 and category 2 and the two sides of the hyperplane is arbitary.

The logical operation described above is performed by a threshold logic unit or TLU. The TLU accepts the pattern vector to be classified, calculates the dot product between the pattern vector and the weight vector, compares the dot product against zero, and classifies the pattern according to the sign of the dot product.

Discriminant Function Development. Given the system discussed above for performing classifications, the outstanding problem in the development of useful pattern classifiers becomes that of finding useful decision surfaces. This can be done, for the nonparametric systems of interest, by a method called training. A training set of patterns whose correct classifications are known is used to develop an effective decision surface.

The members of the training set of objects are presented to the TLU being trained one at a time. The weight vector being trained is initialized arbitrarily. When an incorrect classification is made, the weight vector is altered. The alteration is performed in such a way as to insure that the new weight vector will correctly classify the pattern. This process continues until all the patterns of the training set are correctly classified. If the procedure does not find a weight vector capable of correctly classifying all the members of the training set, then the routine is terminated in order to conserve computer time.

Learning Machine Attributes. The capabilities and performance of learning machines can be described in terms of three principal attributes: recognition, convergence rate, and prediction.

Recognition is the ability of the trained binary pattern classifier to correctly classify the members of its training set. Recognition is 100% for a binary pattern classifier whose decision surface is in the region between two separated clusters. That is, after training is complete for such a case, the TLU can correctly categorize any of the members of the training set.

Convergence rate refers to the rate at which a TLU approaches 100% recognition. Since computer time is an expensive commodity, it is of interest to minimize training time. The training procedures used to find useful TLU's are commonly altered so as to force rapid learning.

Prediction refers to the ability of the TLU to correctly classify unknowns which were not members of the training set. Prediction is the most interesting and potentially useful of the attributes because high predictive ability demonstrates that the TLU has been able to learn something about how to discriminate between the two classes being trained for, and the ability to correctly classify unknown spectra into useful chemical categories is one drive behind all automation of chemical data interpretation. Predictive ability is normally tested by splitting the available data set into two parts - a training set and a prediction set. After training is complete, and without further adjustment of the weight vector, the members of the predictive set are classified and the percentage correct is taken as the predictive ability. Another approach, known as the leave-one-out procedure, involves training a BPC using a training set containing all the data on hand except one member, and then predicting the class of the one unknown after training is complete. When averaged over a number of independent trials, the percentage of unknowns correctly classified is a measure of the predictive ability.

Feedback Feature Selection. After a series of weight vectors have been trained for the same question, then they can be used to perform feedback feature selection. One method that has been used for a number of problems is weight-sign feature selection. Implementation of this method takes advantage of the fact that the exact orientation of a trained weight vector (that is, the relative magnitudes of its components) depends on the initialization used prior to training, the magnitude of the nth component of the pattern vectors, x_n, the feedback training methods employed, the sequence in which the members of the training set are presented to the classifier during training, and several other factors. In other words, the exact orientation of a trained weight vector depends on the details of how it was found. In weight-sign feature selection a pair of weight vectors is developed for the same question but using slightly different approaches, e.g., different initializations. Then the algebraic signs of their components are compared pairwise. When the components of the two weight vectors that both correspond to a particular descriptor disagree in sign, that descriptor is discarded; when the signs agree, the descriptor is retained. The procedure is repeated iteratively until two weight vectors are trained that are in complete agreement for all descriptors that are most useful for a particular classification.

More recently, a new feedback feature selection procedure much superior to the weight-sign method has been developed. The variance feature selection method also takes advantage of the fact that the orientation of a trained weight vector is dependent on how it was developed. Here, a group of weight vectors are trained for a classification problem in a manner designed to exploit these dependencies. The series of weight vectors is then used to rank the descriptors that were most useful in separating the two classes under investigation. The ranking is done by developing an ordered list of the descriptors based on the relative variation of the corresponding weight vector components among the series of trained weight vectors. Then the intrinsic descriptors (those forming the minimal set of descriptors sufficient to effect separation) can be discarded. The variance feature selection method has been applied to a wide variety of problems in our laboratory.

Chemical Applications of Pattern Recognition. Application studies of chemical problems using pattern recognition techniques have been reported in a number of areas (8-14). These are listed in subsets because each general area requires some different approaches and techniques.

(1) Spectral Data Analysis. Elucidation of chemical structure information from spectroscopic data is the area that has received the most attention from those practicing pattern recognition. Studies have been done with mass spectra, infrared spectra, stationary electrode polarograms, gamma-ray spectra, proton and ^{13}C nuclear magnetic resonance spectra.

(2) Materials Science. The classification of materials as to origin or suitability with respect to production specifications has been reported. The data used are generally multi-source data coming from a variety of analytical techniques.

(3) Classification of Complex Mixtures. The identification of petroleum samples by analyzing analytical data by pattern recognition techniques has been reported. Data used for classification in different studies has included gas chromatograms, infrared spectra, fluorescence spectra, trace metals concentrations. A

second example of data analysis of complex mixtures is from the biological mixtures, e.g., serum, are feasible and have been reported.

(4) Modeling of Chemical Experiments. Pattern recognition techniques have been used to model complex chemical systems where the details of the chemical and/or physical interactions were not completely understood, e.g., relative retention of compounds on different chromatographic liquid phases.

(5) Prediction of Properties from Molecular Structure. A number of studies of the application of pattern recognition to the problem of searching for correlations between molecular structure and biological activity have been reported.

Applications of Pattern Recognition to Structure-Activity Relations

Applications to Structure-Activity Relations. Within the last few years reports have begun to appear of work dealing with cluster analysis and pattern recognition applications to drug structure-activity relation studies. A paper by Hansch, Unger, and Forsythe (15) discussed the application of hierachical cluster analysis techniques to the problem of selection of substituents. The data used to represent each drug were the lipophilic π constant, electronic parameters, the approximate steric molar refractivity and molecular weight constants -- physicochemical parameters. A paper by Hill et al. (16) discussed the problem of drug design as approached by using a three-layer perceptron network. Forty-six 1,3-dioxane molecules were used as the data set for training and prediction of perceptrons to determine anticonvulsant activity. Predictive abilities in the range of 68 to 76 percent were reported. A paper by Ting et al. (17) reported correlations between the low resolution mass spectra of sixty-six drugs and their pharmacological activity as sedatives or tranquilizers. This paper was criticized with regard to the set of drugs used in the analysis (18) and with regard to the number of drugs used and their relative similarities (19). Several papers (20-22) have recently appeared reporting studies in which molecules were represented by a list of structural features of the molecules. Adamson and Bush (20) used library searching programs to generate all structural fragments in their data set and represented the drugs by lists of the number of occurences of each substructure in the molecules. Chu (21) used a number of pattern recognition and cluster analysis programs to analyze a set of sixty-six drugs represented by forty-six fragments.

Kowalski and Bender (22) used three pattern recognition classifiers to attempt to classify 200 drugs with respect to activity for the Adenocarcinoma 755 Biological Activity Test. Their paper has been criticized for the choice of the twenty descriptors used (23). Chu et al. (24) reported on the application of pattern recognition and substructural analysis to the problem of investigating the antineopliastic activity of a set of drugs in the experimental mouse brain tumor system. The set of molecules were represented by augmented atom fragments, "heteropath" fragments, and ring fragments. Nearest neighbor and learning machine methods of classification were employed, and it was concluded that these methods could be successfully applied to the problem. Craig and Waite (25) have reported the use of pattern recognition techniques to the prediction of toxicity of organic compounds.

Structure-Activity Studies Using Pattern Recognition

In order to apply pattern recognition techniques to studies of molecular structure-biological activity correlations the data must be taken through a number of individual steps. These are listed in order to show how interrelated the steps become.

(a) Identify data set.

(b) Enter molecular structures. A complete description of the structure of each molecule must be entered into a file.

(c) Generate usable file. A subset of compounds must be selected from the master structure file. This may involve searching of keys for the structures, and will require carrying along an identifying label for each structure.

(d) Descriptor development. The molecular structures stored in a general purpose form (*e.g.*, connection tables) must be decomposed into sets of descriptors. The three general classes are topological, geometrical, and externally generated descriptors.

(e) Form data matrix. The subset of the available descriptors to be used is identified, and a matrix of data is generated. It may be partitioned into a training set and a prediction set.

(f) Prior feature selection. Techniques can be applied to determine which descriptors are expected to be most important.

(g) Discriminant development. The data set is used to develop a discriminant function. After development, the discriminant function can be tested on unknowns to assess predictive ability.

(h) Feedback feature selection. The results of classification can be used to identify the most useful descriptors.

One of the primary prerequisites for a useful general purpose pattern recognition system is a general, data-independent, file management system. A general purpose system has been developed (26) that consists of a set of interactive computer routines known collectively as ADAPT (Automated Data Analysis using Pattern recognition Techniques). This system provides a generalized framework that takes into account the practical considerations inherent in the implementation of the pattern recognition framework shown in Figure 1.

ADAPT Architecture. Figure 1 does not make clear the inherent diversity of the data handling problem. Not only must measurements from the transducer(s) be input, but they must be stored and labelled. Each data point must be given a class designation and indentification number. Class designations must be easily assigned or modified. This ease of definition and redefinition is of utmost importance in the overall data analysis. The source of the data is also important. Sources such as digitized spectra or complex molecular structures would have widely different storage requirements. Since the operations performed on one type of data may bear little similarity to the operations performed on other types of data, a system designed with a high degree of modularity is required. To accomodate these requirements, the ADAPT system is implemented in independent segments. Each segment can execute independently, obtaining all necessary information either from a set of disc storage files or by interaction with the user. This mode of operation offers several advantages, the most obvious of which is a savings in core storage.

The modularity decreases the complexity of the system and provides a means to incorporate additional algorithms into the system at any time. Thus the entire system is adapted to any user's individual requirements since only those overlays which are relevant to the particular problem at hand need be executed. In addition, these routines are relatively inexpensive to use because they do not require large scale facilities for execution. Finally, the system is interactive in the sense that the user directs which manipulations are to be performed upon the data.

ADAPT thus consists of a framework within which an unlimited number of independent segments can be supported. Each segment performs a specific, independent operation ranging from initial input of data to final output of results. The general utility of the system arises from the fact that the user has a large number of options to choose from, and he can conveniently interact with his data set.

Interaction with ADAPT is provided via a Tektronix 4010 CRT terminal. Data is stored in a series of defined files on cartridge discs. This allows fast access and ease of manipulation. Currently, ADAPT consists of approximately 70 defined files which use 2.4 million bytes of storage (one cartridge disc). The ADAPT routine uses approximately 90,000 bytes of core storage for its largest overlay and is currently implemented using a sixteen-bit MODCOMP II/25 computer located in the Department of Chemistry at The Pennsylvania State University.

The segments of the ADAPT system can be broken down into the following list:

(1) File generator, including graphical input of structures
(2) Class maker
(3) Three-dimensional model builder
(4) Descriptor developer

(5) Collator
(6) Preprocessor
(7) Prior feature selector
(8) Discriminant developer
(9) Feedback feature selector

(1) File Generator. The library of drugs to be studied is entered through the file generator routine. Structures are input by drawing them in two dimensions on the screen of an interactive graphics terminal under the control of a general structural input routine, UDRAW, which has been fully described elsewhere (27). A molecule's structure, along with corresponding pharmacological data, is entered into a disc resident permanent file. Information saved for future use includes a compressed connection table, ring information, a list of reported activities, the two-dimensional coordinates of the atoms when entered (for possible redrawing of the structures later), an identification number, and the chemical name of the compound. In addition to generation, the file can be altered by making changes to information stored for a drug, a drug can be entirely deleted from the file, or any file member can be displayed. A selection of recallable molecular backbones can be stored for more convenient entry of series of structurally related compounds. These structures can then be made to appear upon the initial UDRAW sketch pad and a complete molecule can be built up starting from this backbone. This allows the user to input a series of structurally similar compounds without redrawing the base structure each time. The routine that oversees structure input and file generation can maintain a file of 1000 structures and associated auxiliary information.

The first structure file now stored in the system consists of approximately one thousand central nervous system agents taken from the literature (28). Among the biological activity classes reported there are analgesics, anticonvulsants, depressants, hypnotics, relaxants, sedatives, stimulants, and tranquilizers; there are approximately forty classes altogether, many of which overlap.

The second file of molecular structures currently resident on the ADAPT disc file consists of 184 5,5-disubstituted barbiturates taken from a reference volume (29). A study using this data set will be discussed in a later section.

The third file contains approximately 500 compounds comprising an olfaction data set taken from Amoore (30). Molecules reported to have musk, camphor, mint, ether, floral, pungent, and putrid odors are present. This data set is being used in studies of the relation between molecular structure and odor quality.

The fourth file consists of a set of molecules comprising an olfaction data set taken in a study of trigeminal detection of compounds. These compounds are being employed in a study of the similarities and differences observed in trigeminal as opposed to olfactory detection of chemicals by humans.

(2) Class Maker. The class maker routine is used to access the library file and to create sets of library members that satisfy queries entered by the user. Thus, the set of all file members which have been reported to be sedatives can be formed into an active data set. This routine is used to generate classes of structures to be used as data sets for the development of discriminants by another section of ADAPT.

When the property being sought is known quantiatively, the data set is assembled in increasing sequence. Then a series of discriminants can be trained for different threshold cutoffs between the active and inactive classes without moving any data but only by reallocating class memberships.

(3) Three-Dimensional Model Builder. The three-dimensional molecular model builder routine is used to derive information on the spacial conformation of molecules. A molecule can be viewed as a collection of particles held together by simple harmonic or elastic forces. These forces can be defined by potential energy functions whose terms are the atom coordinates of the molecule. This function can then be minimized to obtain a strain-free three-dimensional model of the molecule. Geometric parameters can then be extracted. A wealth of information already exists describing the procedures and results of several different molecular mechanics algorithms (31,32). Therefore, finding and implementing an algorithm to model sets of molecules is a relatively straightforward procedure. A modified version of the molecular mechanics routine described by Wipke, et al., (33-35) has been developed and interfaced to the ADAPT system so that geometric descriptors can be derived from the resulting molecular structure.

The molecular mechanics routine, MOLMEC, used in conjunction with the ADAPT system is highly interactive and relies on graphical input and output. A graphics unit is also supported and is utilized by MOLMEC for displaying the molecule being modelled.

The structure input section of MOLMEC has been designed to allow the user to either read the molecule's connection table from ADAPT's disc files or else accept the structure from the CRT via UDRAW (27). Thus, MOLMEC can be used independently of the ADAPT system. Once the molecule has been entered, control branches to the interactive section where the user can direct the different phases of modelling as well as monitor the results.

In the strain minimization section, the atom coordinates are systematically altered until a minimum is found in the strain or potential energy function. The actual strain function used in MOLMEC is:

$$E_{strain} = E_{bond} + E_{angle} + E_{torsion} + E_{non\text{-}bond} + E_{stereo}$$

The first four terms of the function are commonly found in all molecular mechanics strain functions and are modified Hooke's Law functions. The last term of the function has been added to assure the proper stereochemistry about an asymmetric atom.

The actual minimization of the function is best accomplished by some type of nonlinear programming method (e.g., steepest descent). In MOLMEC, an adaptive pattern search routine (36) is used because it does not require analytical derivatives. The amount of time necessary to obtain good molecular models depends upon the number of atoms in the molecule, the initial strain of the molecule, and the degrees of freedom in the structure. If a small molecule is being modelled, only one pass through the minimization section may be sufficient to obtain a good structure. However, this is seldom the case. Usually, the molecules are rather large and require several passes. The actual amount of time per pass is limited by a cutoff parameter so that the user may analyze the progress of the modelling at different intervals.

The graphics interaction section of MOLMEC contains routines capable of rotating and aligning the molecule into any desired position. Since the graphics unit is only a two-dimensional screen, rotation is essential to obtain a good view of the structure. Furthermore, these routines are useful in locating atoms trapped in local minima. If such an atom is found, the user can move the trapped atom to a new position by a MOVE routine found in the graphics section. Naturally, if the structure is altered the molecule should be passed through the minimization routine at least once more.

When the molecule is finally in a low strain energy conformation, the molecular parameters can be either listed on an output device, or else the structure's coordinate matrix can be stored on a disc file for further processing.

An automatic version of MOLMEC has also been developed so that large molecular data sets can be modelled without continuous supervision. The program consists on an input section, which reads the molecule's connection table and present coordinate matrix from the ADAPT files, a minimization section with all output suppressed, and a section which stores the final coordinate matrix. Good models can easily be obtained in this manner. However, before the coordinate matrices can be used for calculating descriptors, the structures are reviewed to make sure that the molecules are in acceptable conformations. Once modelling is complete, geometric descriptors can be derived.

Descriptors currently being used include the absolute or relative magnitudes of the principal moments of inertia of the molecule, the presence or absence of particular spacial arrangements of atoms which have been called pharmacophores, and the molecular volume.

(4) Descriptor Developer. The next step in studies of structure-activity relations is the development of descriptors for the molecules contained in the active data set. This subject has been

discussed in a recent publication (37). Descriptors belong to two general classes: topological and geometrical. Topological descriptors are derived from the topological representation of a compound -- the connection table. Geometrical descriptors are derived from the three-dimensional model of the molecule. The individual descriptors that have been used in reported studies are described in the following paragraphs.

(a) Atom and bond descriptors -- Fragment descriptors. Atom descriptors include the number of C, N, O, S, P, F, Cl, Br, I atoms in the structure. Numbers of bonds of each type are also generated. Both atom and bond descriptors are developed directly from the stored connection table.

(b) Substructure Descriptors. Searching the molecule for the presence of larger fragments provides an alternative method for generating descriptors. If the substructure is found in the molecule, the descriptor can be given a value of one. Otherwise, it has a value of zero. Therefore, to generate substructure descriptors for a given molecular data set, two things are needed: a substructure searching algorithm and a library of appropriate substructures.

Algorithms for substructure searching fall into two general categories. The first, atom-by-atom searching, is the easiest to implement on a digital computer because it simply matches the structure and substructure atoms and associated bonds one at a time using all possible combinations. However, for large structures and substructures the time required for a single search becomes prohibitive because of the number of possible combinations increases factorially.

The second category utilizes set reduction techniques to accomplish the substructure search, and factorial calculations are not involved. Although they are more complex than atom-by-atom searching techniques, algorithms implementing set reduction are very attractive because of their searching speed. Several different algorithms have been described which use set reduction (38-40). In the ADAPT system, a variation of the techniques described by Sussenguth (38) is used for generating substructure descriptors. The modifications allow for greater substructure specificity, a wider variety of substructure types, and numeric instead of binary searches. A discussion of the changes made in the Sussenguth's algorithm has been reported (41) and will not be detailed here.

The problem of creating a substructure library is not as easy to solve as obtaining a good substructure searching algorithm. One approach to this problem involves the systematic combing of the basic atom and bond fragments into substructures. However, the final number of substructures generated in this manner would be totally unmanageable. The discrimination between usable and useless substructures would require some type of pattern recognition system, and this approach is not feasible. A more workable approach to the problem is to study the data set of molecules under investigation and allow the chemist to decide on a collection of substructures to be applied to the data set. The ADAPT system utilizes this second method to generate a substructure library. A set of substructure descriptors can now be generated.

Two types of searches are possible. For a general search, a match is made if the indicated substructure is located anywhere in the molecule; all ring information is ignored. However, during a specific search, ring information is taken into consideration. Therefore, if the substructure is not specified to be in a ring, it cannot possibly be matched to a molecular fragment that is contained in a ring system.

The actual information contained in any one substructural descriptor depends highly upon the judgement of the person selecting the substructure library. In some applications, good descriptors can be obtained immediately because sufficient *a priori* knowledge exists. However, in other cases, a trial-and-error procedure may be warranted where a large number of possible substructures are generated and poor descriptors are eliminated by some prescreening criterion. In general, substructure descriptors serve a very important purpose in that they restore a portion of the structural information lost in the atom and bond fragmentation. Nevertheless, considerable structural information is still missing.

(c) Environment Descriptors. The description of structures using fragment and substructure descriptors indicate the components of a molecule. However, the manner in which these individual parts are connected is not described. Environment descriptors take into account how different areas of a molecule fit together and provide a measure of the "environment" in which a single atom fragment finds itself.

The environment descriptor describes the fragment's surroundings by including its first and second nearest neighbors and their bonds into a single parameter which reflects the atom and bond types connected to it. There may be more than one identical fragment in a molecule but they do not necessarily belong to the same functional group. For example, the fragment, -C-, is found once in both structures A and B below, but twice in structure C:

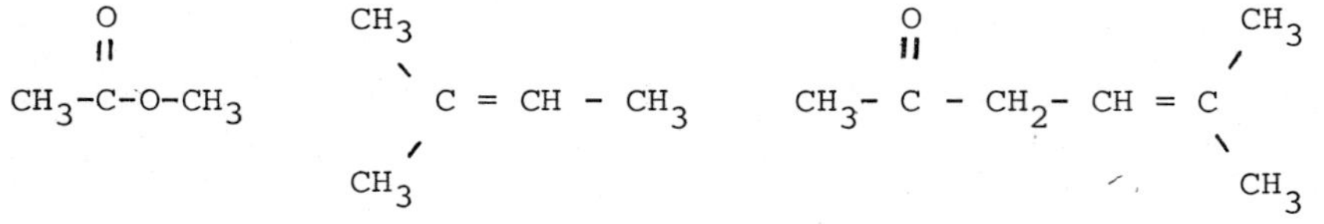

```
  O            CH3                          O                 CH3
  ||              \                         ||               /
CH3-C-O-CH3        C = CH - CH3      CH3- C - CH2- CH = C
                  /                                          \
               CH3                                            CH3

   (A)                (B)                      (C)
```

Obviously, the environment seen by this fragment would be different in each of the three cases. Of course, this difference is dependent upon the definition incorporated to calculate the environment descriptor. In the ADAPT system, the three forms most often used are: bond environment descriptors (BED), weighted environment descriptors (WED), and augmented environment descriptors (AED).

The procedure used to calculate these three parameters for a particular environment fragment is as follows:

(1) Assign arbitrary values to each type of atom and bond. The values already employed in the connection table will suffice.

(2) For "BED", sum the number of bonds connected to the fragment's first and second nearest neighbor.

(3) For "WED", sum the values assigned to each bond type instead of merely counting the bonds.

(4) For "AED", sum the product of the bond's assigned value and the assigned values for the two atoms which form the bond.

The BED, WED, and AED values for the fragment and structures given above are as follows: for structure A, BED = 5, WED = 6, AED = 11; for structure B, BED = 5, WED = 6, AED = 6; for structure C, BED = 12, WED = 15, AED = 17.

Since there may be more than one fragment present, the environment descriptor indicates the sum of all the environments for a given fragment. This feature makes them useful when used in conjunction with substructure descriptors. The substructure descriptors indicate the number of times a particular fragment is found in the molecule and the environment descriptors indicate the context in which the fragment is found.

The routine that generates the environment descriptors must have access to the file of molecular structures and to the atom centered fragment library which is constructed by the user. The actual calculation of the environment descriptors proceeds extremely rapidly since both the fragment location and necessary calculations are easily done by a computer.

The concept of the environment is not limited to connectivities, but could take into account electron densities, bond distances, electronegativities, or other physical parameters. This can be done by replacing the values assigned in step one by the desired parameters. In this manner, more informative descriptors may be obtained.

Use of the environment descriptors may reveal relations which are not particularly obvious. Note that both structures A and B have the same BED and WED values. These structures, which at first glance appear quite different, do indeed have these parameters in common. However, when one takes into account the type of atoms connected to these bonds the difference becomes apparent. Such relationships may or may not prove significant. Their ultimate utility depends on the type of environment measure, the molecule being coded, and the problem being attacked.

(d) Geometric Descriptors. Geometric descriptors are derived from the three-dimensional configuration as generated by MOLMEC. Presently, two basic types of geometric descriptors are calculated from the molecular structures. The three principal axes of the molecule form the basis for the first type of geometric descriptor.

Since the orientation of the original molecule in space is

essentially random, the radii must be sorted in some manner. This is done by arbitrarily assigning X to the longest radius, Y to the second longest radius, and Z to the shortest radius. Once sorted, the three ratios, X/Y; X/Z and Y/Z, are also calculated. Due to their small values, all of the radii are multiplied by some constant scaling factor to prevent loss of information during truncation. These six geometric parameters are then used as new descriptors and constitute the first set of geometric descriptors.

The van der Waals volume of a molecule is the other type of geometric descriptor generated in the ADAPT system. Before this calculation can be done, the bond distances and the van der Waals radii of the atoms must be known. The bond distances are easily obtained from the molecular modelling results. For the van der Waals radii, an article published by A. Bondi (42) was consulted. The volume occupied by an atom is taken as that of a sphere with radius equal to the van der Waals radius of the atom minus the volume of overlap with adjacent atoms. The overlap volumes are calculated from standard spherical geometry formulas. The actual volume is not found for two reasons: the assumption of sphere and spherical segments is not totally correct, and the radii used were selected as being the "best" values from a large collection of data using an empirical selection method. The total molecular volume for the molecule is taken as the sum of the contributions for each atom found as described above. The volume contributions of attached hydrogens are also included in the calculation of the total volume.

In order to make the routine more versatile, the option of either using standard bond distances or modelled bond distances is included. Since MOLMEC uses the standard bond distances to determine a low strain geometry, it is not surprising that for a well modelled data set, the molecular volumes calculated using the two different bond distances are very similar. However, discrepancies can arise when the molecule contains rings of five or fewer atoms which cause a large amount of bond strain. The volumes are initially calculated in units of cubic Angstroms per atom but are then converted to units of cc per mole. The molecular volume can then be used as another geometric descriptor.

Each geometric descriptor contains some information about the molecule. The radii and ratios describe the general shape of the molecule which may be very important in systems where receptor sites are involved. However, this is only a relative shape since the model obtained is for the molecule in a vacuum: in some environments, the molecule's shape will change, especially if long chains are present. On the other hand, the molecular volume is essentially constant regardless of how the molecule is bent. However, like any other descriptor, the actual value of any geometric descriptor depends upon the specific application in which it is used.

(5) Collator. The collator routine is used to select which of the

available descriptors will be included in the data set to be passed to other parts of ADAPT. The experimenter has complete flexibility in deciding which data set or subset to use and how to structure problems when they are to be passed to the prior feature selection algorithms of the discriminant development algorithms. This routine is used to select first one subset of the available descriptors to be used for discriminant development, and then on subsequent trials other subsets of descriptors. Thus, overall performance of the system can be evaluated with respect to which descriptors are being included in the analysis.

(6) Preprocessor. The preprocessor routine accepts the raw descriptors developed by the descriptor development routines and performs the desired preprocessing necessary for further processing. One example of such preprocessing is autoscaling, where each descriptor over a data set is altered so that the mean is zero and the standard deviation is unity. The statistics literature calls this procedure standardizing the variables.

(7) Prior Feature Selection. After a set of drugs have been formed into a labelled data set ready for presentation to the discriminant developer, it is desirable to submit it to feature selection if possible. One method for selecting the descriptors expected to be most useful has been the use of the well-known Fisher ratio (*e.g.*, 21). A number of other statistically based methods suggest themselves, but they mostly require making the assumption that the best, *i.e.*, most separating, descriptors identified *one at a time* will also be the best set of descriptors. This assumption is rarely valid.

In the studies performed to date, we have usually tried to select subsets of descriptors in as wise a manner as we could devise; we have relied on being able to investigate a large enough number of subsets of descriptors to feel reasonably confident that we have found good descriptor sets.

Feature selection is performed as an integral part of stepwise descriminant analysis such as that implemented in the BMD (43) package as BMDO7M. This will be discussed later in the section on discriminant development and feedback feature selection.

(8) Discriminant Developer. The discriminant developer accepts the set of data generated by the previous sections of ADAPT and attempts to develop discriminant functions capable of correctly classifying the data. The development of such discriminants can proceed through the use of (a) error correction feedback learning machines, (b) interactive least squares development of linear discriminant function, (c) other parametric and nonparametric routines. The error correction feedback training method has been used in the studies on barbiturates to be described in the following section of this article. The iterative least squares development method was developed several years ago in this laboratory (44) and has been

interfaced into ADAPT.

(9) Feedback Feature Selection. In many chemical applications of pattern recognition a set of data is coded using more descriptors than are necessary to correctly classify the members. However, the necessary and unnecessary descriptors cannot usually be identified *a priori*. (When they can, this is obviously the method of choice.) Therefore, feature selection must often be approached from a systems viewpoint, whereby the results of classification are used to try to identify the minimal set of necessary descriptors. This approach is shown by the feedback loop in Figure 2.

An early approach to feedback feature selection was weight-sign feature selection. Here, two weight vectors, initialized with each component equal to +1 or -1, respectively, were developed using error correction feedback training with identical training sets. A component by component comparison was made between the two trained weight vectors, and those descriptors corresponding to weight vector components with sign disagreements were discarded. This method was shown to be effective for some classes of data in several studies. The variance feature selection method, described earlier, has been incorporated into ADAPT and has been used effectively on several types of data. The variance method allows rapid extraction of features responsible for linear seperability. It is much superior to the weight-sign method in terms of speed and reliability.

Barbiturate Study

The set of compounds used in the present study consists of 160 5,5'-substituted barbiturates selected from a standard reference (29). These compounds range in molecular weight from 172 to 276 and have duration times ranging from 10 minutes to 600 minutes. The method of administration was either intraperitoneal or subcutaneous, using mice, rats, or rabbits as test animals.

The compounds were grouped into classes according to the duration of depressant effect. These classes were formed by dividing the duration time expressed in minutes by ten. The resulting class designation was rounded up if the remainder was five or greater, and down otherwise. Thus a compound whose duration time was 227 minutes would be placed in class 23, whereas a compound having a duration time of 223 minutes would be placed into class 22. Compounds with a duration greater than 650 minutes were placed into class 65. This resulted in a total of 65 different classes which are distributed as shown in Fig. 3.

Three types of descriptors were employed for these studies; numeric fragment descriptors, substructural descriptors, and environmental descriptors. The descriptors were generated using the automated descriptor packages described previously. A list of the initial set of descriptors used is given in Table 1. Each descriptor is contained in a minimum of 20% of the structures. In no case

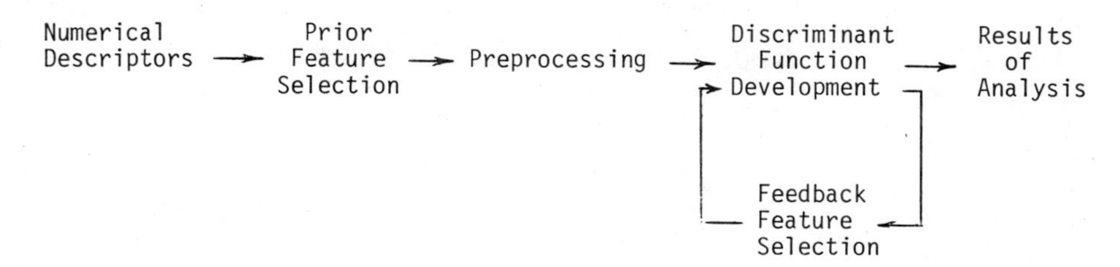

Figure 2. Basic pattern recognition system for studies of structure–activity relationships

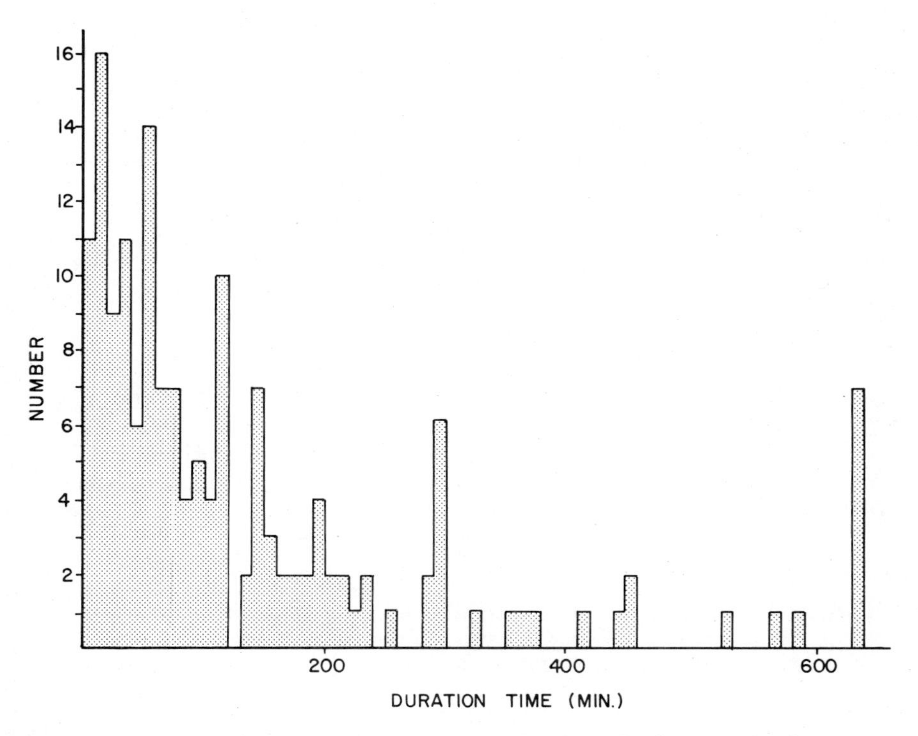

Figure 3. Histogram of barbiturate duration times for the drugs in the data set

TABLE I. Molecular Structure Descriptors for the Barbiturate Data Set

ATOM AND BOND DESCRIPTORS

1	Number of atoms	2	Number of bonds
3	Number of Carbon atoms	4	Number of Nitrogen atoms
5	Number of Oxygen atoms	6	Number of single bonds
7	Number of double bonds	8	Length[a]

ENVIRONMENT DESCRIPTORS

	Atom Centered Fragment	General[b]	Cyclic
9 - 11	CH_3-	1, 2, 3	
12 - 14	-CH_2-	1, 2, 3	
15 - 17	-CH-	1, 2, 3	
18 - 23	-C -	1, 2, 3	1, 2, 3
24 - 26	O =	1, 2, 3	
27 - 29	-HC =	1, 2, 3	
30 - 35	>C =	1, 2, 3	1, 2, 3

SUBSTRUCTURAL DESCRIPTORS

36	CH_3CH_2-	37	-$CH_2(CH_3)CH$ -	38	CH_3-
39	-CH_2-	40	-CH_2CH_2-	41	$CH_3CH_2CH_2$-
42	-CH-	43	-HC =		

[a]Length = 4*(Number of single bonds) + 2*(Number of double bonds)

[b]1 = BED, 2 = WED, 3 = AED

does any one descriptor, or any binary combination of descriptors, contain sufficient information to successfully classify the data.

Thus the active data set consists of 160 compounds each coded with 43 descriptors. Preprocessing of the raw data prior to classification consisted of autoscaling so that each descriptor had an average of zero and a standard deviation of 127. This allowed the data to be truncated to integer values with a negligible loss of precision. (Loss of precision is known to be negligible as recalculation after truncation yielded a standard deviation of 127 and a mean of 0 ± 0.17.) Net retention of information was assured by testing the predictive ability for each descriptor before and after preprocessing. A value of 250 was used for X_{n+1} because it provided fast training and high predictive ability.

Since the data were collected from a series of studies on different animals, at different laboratories, it is not unreasonable to expect the classifications to differ. It was therefore felt that an error range would take into account the variations due to different classification methods. Thus, any one classifier will develop a rule which answers the question, "Is the duration time less than x minutes?", where there is a deadzone of several minutes around this level. Thus, to test for discrimination ability at a threshold level of 100 minutes using a thirty minute deadzone, all members from classes 1 through 10 would constitute one category, and all members from 14 through 65 would constitute the other category.

The linear learning machine was used to develop discriminant functions which bisect the data with as many different thresholds as possible, obtaining 100% recognition ability for each range. Attempts at such discrimination were accomplished using first a fifty, and later a thirty, minute error range.

To generate a preliminary estimate of the clustering and self consistency of the data the following experiment was done. Five training set/prediction sets were chosen with seven compounds in each prediction set and the remaining compounds in each training set. The overall data set is divisible into halves by 59 thresholds using 50 minute error ranges. All five training sets were used to develop independent discriminants at each of the 59 thresholds. These discriminants were then used to predict the seven unknowns in the respective prediction set. The class assignments were made by examining the sequence of responses produced by the 59 predictions; if only one change from answers of "greater than" to "less than" occurred, this point was taken as the predicted duration time. If there were several changes in predicted response, then the predicted duration time was taken as 30 minutes greater than the shortest duration time indicated by the first change in response. When this procedure was used, 19 of the 35 unknowns were classified as having duration times within 20 minutes of the actual value and 31 were classified as having duration times within 50 minutes of the actual value. The duration times

TABLE II. Final Sets of Molecular Structure Descriptors Supporting Linear Discriminant Functions at Thresholds I and II.

THRESHOLD I		THRESHOLD II	
ATOM AND BOND DESCRIPTORS		ATOM AND BOND DESCRIPTORS	
Number of Oxygen atoms		Number of Oxygen atoms	
Number of double bonds		Number of single bonds	
SUBSTRUCTURAL DESCRIPTORS	ENVIRONMENT DESCRIPTORS[a]	SUBSTRUCTURAL DESCRIPTORS	ENVIRONMENT DESCRIPTORS[a]
CH_3-	CH_3-(G,2)	CH_3-	-HC-(G,1)
-CH_2-	-CH-(G,1)	-CH_2CH_2-	-HC=(G,1)
-CH_2CH_2-	-HC=(G,2)	-C-	>C=(G,3) (C,3)
CH_3CH_2-	>C=(G,3) (C,1)	-CH_2CH (CH_3)-	
Average Predictive Ability[b]	93.8%	Average Predictive Ability[b]	94.9%

[a]G = General search, C = Cyclic search, 1 = BED, 2 = WED, 3 = AED

[b]Predictive ability measured using leave one out procedure

of only four compounds were in error by more than 50 minute error range used for each threshold. Thus this preliminary experiment showed that a set of linear classifiers working in concert could predict the duration times of the compounds in the data set reasonably accurately. Similar results were obtained for the 61 possible thresholds developed using a 30 minute error range.

In order to gain a better insight into these relationships two thresholds were subject to exhaustive feature selection. The threshold I data includes classes 1 through 10 and 14 through 65. The threshold II data includes classes 1 through 24 and 28 through

65. The results of the selection process and a predictive ability estimation is reported in Table II.

Through application of the variance feature selection method, a set of features responsible for the separability of the data were found. Removing any of these descriptors results in the loss of linear separability. Therefore, the descriptors selected constitute a minimum set capable of supporting the relationship within the data. The predictive ability, estimated by the leave one out procedure (45), indicated that these features were capable of providing accurate information concerning the duration of barbiturate activity. Thus, it is clear that a relationship is present which is readily identified using the ADAPT system.

Further investigations using this data set have uncovered several interesting correlations. Details of the experimental results are reported elsewhere (46). What has been sought for here is a clear demonstration of the utility of ADAPT in ellucidating relations within a large body of data. Note that feature selection of the two specific thresholds was easily accomplished as was initial development of discriminants for 61 different classes. Clearly such studies would be inconvenient without the degree of organization provided by automation of the descriptive, storage, and pattern recognition techniques. The ADAPT system has consistently shown high utility in several areas and promises to continue to aid in the application of pattern recognition to problems in chemistry.

Literature Cited

1. Minsky, Marvin, Proc. IEEE, 49, 8 (1961).
2. Solomonoff, R.J., Proc. IEEE, 54, 1687 (1966).
3. Rosen, C.A., Science, 156, 38 (1967).
4. Nagy, George, Proc. IEEE, 56, 836 (1968)
5. Levine, M.D., Proc. IEEE, 57, 1391 (1969).
6. Nilsson, N.J., Learning Machines, McGraw-Hill Book Co., New York, 1965.
7. Tou, J.T. and Gonzalez, R.C., Pattern Recognition Principles, Addison-Wesley Publishing Co., Reading, Mass., 1974.
8. Kowalski B.R. and Bender, C.F., Jour. Amer. Chem. Soc., 94 5632 (1972); 95, 686 (1973).
9. Isenhour, T.L., Kowalski, B.R., Jurs, P.C., Crit. Rev. Anal. Chem., 4, 1 (1974).
10. Kowalski, B.R., "Pattern Recognition in Chemical Research," in Computers in Chemical and Biochemical Research, Vol. 2, C.E. Klopfenstein and C.L. Wilkins, Eds., Academic Press, New York, 1974.
11. Jurs, P.C. and Isenhour, T.L., Chemical Applications of Pattern Recognition, Wiley-Interscience, New York, 1975.
12. Kowalski, B.R. and Bender, C.F., Naturwissenschaften, 62, 10 (1975).

13. Kowalski, B.R., Anal. Chem., 47, 1152A (1975).
14. Jurs, P.C., Proceedings of the Workshop on Chemical Applications of Pattern Recognition, Washington, D.C., May 1975.
15. Hansch, C., Unger, S.H., Forsythe, A.B., Jour. Med. Chem., 16, 1217 (1973).
16. Hiller, S.A., et al., Comp. Biomed. Res., 6, 411 (1973).
17. Ting, K.-L.H., et al., Science, 180, 417 (1973).
18. Perrin, C.L., Science, 183, 551 (1974).
19. Clerc, J.T., Naegeli, P., Seibl, J., Chimia, 27, 639 (1973).
20. Adamson, G.W. and Bush, J.A., Nature, 248, 406 (1974).
21. Chu, K.C., Anal. Chem., 46, 1181 (1974).
22. Kowalski, B.R. and Bender, C.F., Jour. Amer. Chem. Soc., 96, 916 (1974).
23. Unger, S.H., Cancer Chem. Rpts., Part 2, 4(4), 45 (1974).
24. Chu, K.C., et al., Jour. Med. Chem., 18, 639 (1975).
25. Craig, P.N. and Waite, J.H., Analysis and Trial Application of Correlation Methodologies for Predicting Toxicity of Organic Chemicals, EPA Office of Toxic Substances, 1976.
26. Stuper, A.J. and Jurs, P.C., Jour. Chem. Infor. Comp. Sci., 16, 99 (1976)
27. Brugger, W.E. and Jurs, P.C., Anal. Chem., 47, 781 (1975).
28. Usdin, E. and Efron, D.H., Psychotropic Drugs and Related Compounds, 2nd ed., DHEW Publication No. (HSM) 72-9074, 1972.
29. Doran, W.J., Medicinal Chemistry, Vol. IV, John Wiley and Sons, New York, 1959.
30. Amoore, J.E., Molecular Basis of Odor, Thomas, Springfield, Ill., 1970.
31. Engler, E.M., Andose, J.D., Schleyer, P. von R., Jour. Amer. Chem. Soc., 95, 8005 (1973).
32. Williams, J.E., Strang, P.J., Schleyer, P. von R., Ann. Rev. Phys. Chem., 19, 531 (1968).
33. Wipke, W.T., Dyott, T.M., Verbalis, J.G., Abstract, 161st American Chemical Society National Meeting, Los Angeles, CA, March 1971.
34. Wipke, W.T., Gund, P., Verbalis, J.G., Dyott, T.M., Abstract, 162nd American Chemical Society National Meeting, Washington, DC, September 1971.
35. Wipke, W.T., Gund, P., Dyott, T.M., Verbalis, J.G., unpublished manuscript.
36. Buffa, E.S. and Taubert, W.H., "Production-Inventory Systems, Planning and Control," Rev. Ed., R.D. Irwin, Inc., Homewood, Ill., 1972.
37. Brugger, W.E., Stuper, A.J., Jurs, P.C., Jour. Chem. Infor. Comp. Sci., 16, 105 (1976).
38. Sussenguth, E.H., Jr., Jour. Chem. Soc., 5, 36 (1965).
39. Ming, T.-K. and Tauber, S.J., Jour. Chem. Doc., 11, 47 (1971).
40. Figeras, J., Jour. Chem. Doc., 12, 237 (1972).
41. Zander, G.S. and Jurs, P.C., Anal. Chem., 47, 1562 (1975).
42. Bondi, A., Jour. Phys. Chem., 68, 441 (1964).

43. Dixon, W.J., Ed., BMD-Biomedical Computer Programs, 3rd Ed., Univ. of Calif. Press, Berkeley, CA, 1973.
44. Pietroantonio, L. and Jurs, P.C., Pattern Recog., 4, 391 (1972).
45. Lachenbruch, P.A. and Miche, R.M., Technometrics, 10, 1 (1968).
46. Stuper, A.J. and Jurs, P.C., submitted for publication.

10

Enthalpy-Entropy Compensation: An Example of the Misuse of Least Squares and Correlation Analysis

R. R. KRUG
Chevron Research Co., Richmond, CA

W. G. HUNTER
Statistics Department and Engineering Experiment Station,
University of Wisconsin, Madison, WI 53706

R. A. GRIEGER-BLOCK
Chemical Engineering Department, University of Wisconsin, Madison, WI 53706

Whether or not a linear functional relationship exists between reaction or equilibrium enthalpies and entropies has been the subject of chemical investigations for many years. Hinshelwood collected lots of data during the early years of modern kinetic theory to probe for possible functional dependencies between the Arrhenius parameters (1-3). Many of these and subsequent experimental investigations have led to findings that estimated enthalpies varied linearly with estimated entropies. Many chemical theories have been proposed to explain, in chemical terms, why such linear correlations should occur. Linear enthalpy-entropy compensation is now widely accepted as occurring because of chemical factors and is mentioned in many standard chemistry tests (4-8).

In the past few decades, first chemists (9-16) and later statisticians (17-24) have begun to doubt that all enthalpy-entropy compensations arise as a result of chemical factors alone. In particular as the compensation temperature, the slope of a compensation line in ΔH-ΔS coordinates, approached the range of experimental temperatures, the chemical causality of such correlations was questioned.

The debate over which observed correlations were caused by chemical factors and which were caused by nonchemical factors (i.e. data handling artifacts that result from the propagation of errors) apparently has not been adequately resolved to date because enthalpy-entropy compensations are still reported and justified merely by the significance of the estimated correlation coefficient. In this article we summarize and generalize our earlier results (25-27) that indicate that the significance of an estimated correlation coefficient in the enthalpy-entropy plane is _not_ justification for the detection of a chemically caused compensation, but the significance of an estimated correlation coefficient in the enthalpy-free energy plane with estimates evaluated at the harmonic mean of the experimental temperatures is strong justification for the detection of a chemical effect. This conclusion results from the fact that there is a linear statistical compensation effect that is confounded with whatever chemical compensation that might be detected in the enthalpy-entropy plane. We also present the regression algorithm for the estimation of the chemical compensation temperature from an observed correlation in ΔH-$\Delta G_{T_{hm}}$ coordinates.

Chemical Theory

The rigorous thermodynamic and statistical mechanical arguments of Laidler (28), Hammett (5), Leffler (7,16), and Ritchie and Sager (29) all suggest a generally nonlinear functional relationship between enthalpies and entropies. To illustrate this result, we call upon the statistical mechanical definitions used by Ritchie and Sager.

The entropy of a system and the enthalpy of a system can be written in terms of the sums of energy states that the system occupies.

$$S = R\ln\Sigma g_i \exp(-\varepsilon_i/kT) + R\,\frac{\Sigma g_i(\varepsilon_i/kT)\exp(-\varepsilon_i/kT)}{\Sigma g_i \exp(-\varepsilon_i/kT)}$$

$$H = RT\,\frac{\Sigma g_i(\varepsilon_i/kT)\exp(-\varepsilon_i/kT)}{\Sigma g_i \exp(-\varepsilon_i/kT)}$$

If we take as the system a chemical plus its solvent undergoing reaction or equilibrium, two systematic variations that will cause coincidental variations in enthalpies and entropies are homologous variations of either solvent composition (e.g., from polar to nonpolar) or substituents (e.g., from electron releasing to electron withdrawing). Passing through the homologous series the energy states occupied by the system will vary in a systematic manner. Since the same energy states define all thermodynamic functions of the system, the thermodynamic parameters (including enthalpy and entropy) will also vary in a systematic manner such that a plot of enthalpy versus entropy, say, would reveal a systematic variation. That a systematic variation should be linear is not obvious from the definitions, however. We may assume that if a resultant variation is over a special region or is sufficiently short, the plotted variation may appear to be linear. It is important to note that this would be a linear segment of an otherwise nonlinear function.

Such a linear variation of enthalpy-enthropy pairs ($\underline{\Delta H}$,$\underline{\Delta S}$) is generally summarized as

$$\underline{\Delta H} = \beta\underline{\Delta S} + \Delta G_{\beta}$$

where the slope, β, has the dimension of temperature and is alternately called the compensation temperature, isokinetic temperature or isoequilibrium temperature depending on whether the thermodynamic parameters were estimated from kinetic or equilibrium data. The physical significance of the compensation temperature is that at this temperature a variation in enthalpy is entirely compensated for by a corresponding variation in entropy such that the free energy is a constant. To be consistent with the Gibbs equation, the intercept of such a linear relationship is the free energy at the compensation temperature, ΔG_{β} (9,13).

Statistical Theory

Historically, compensation temperatures have been determined by least squares (or best graphical fit, which is essentially least squares without the computational rigor) and the goodness of fit has been justified by the high significance of the estimated correlation coefficients between the enthalpy and entropy estimates. Both of these procedures are incorrect and, particularly in this case, often lead to grossly incorrect results. It is important to remember that the enthalpy-entropy data pairs are actually estimates ($\underline{\Delta\hat{H}}$,$\underline{\Delta\hat{S}}$) not original data that can be treated as either independent or as being relatively free from error as might be rationalized for original laboratory data, for example, kinetic rate constants-temperature data ($\underline{k}$,$\underline{T}$) or chemical equilibrium constants-temperature data ($\underline{K}$,$\underline{T}$).

The enthalpy and entropy estimates, $\underline{\Delta\hat{H}}$ and $\underline{\Delta\hat{S}}$, both contain uncertainty, and hence least squares is an improper technique for regression of a functional dependence of one on the other. What is worse, actually, is that these estimates are highly correlated with one another due to their functional relationships with the kinetic or equilibrium constants and the experimental ranges over which the data were sampled. Hence a correlation analysis might detect a significant correlation that results from these computations as data handling artifacts, even in the absence of any chemical effect. The statistical and chemical compensations need not be hopelessly confounded, however, because the scientist has knowledge of both the chemical identities and his choice of experimental sampling points prior to analysis.

Using the fundamental definitions of chemical kinetics and regression analysis, we will now show (1) that enthalpy-entropy estimates are highly correlated, (2) that the statistical compensation equation is functionally identical to the chemical compensation equation, (3) how to separate the chemical from the statistical effect, and (4) how to estimate the chemical compensation temperature and its (1-α) confidence interval from kinetic or equilibrium data. To avoid redundancy, we will restrict this discussion to the case of kinetic data, but for completeness we will include the computational details for equilibrium data as well in the Regression Algorithm.

In this discussion, we must make the usual assumptions that errors associated with the dependent variable, the logarithm of the kinetic observations, $\underline{y_i} = \underline{\ln}\ \underline{k_i}$, are normally and independently distributed with zero mean and constant variance, $\underline{\varepsilon} \sim NID(\underline{0},\sigma^2)$, and that the independent variable, the inverse experimental temperatures, $\underline{x_i} = \underline{1/T_i}$, have no uncertainty. That is, in practice the experimental temperatures are determined with much greater precision and accuracy than are the rate constants. To formalize this analysis we consider data taken at $1 \le i \le n$ temperatures for $1 \le j \le m$ members of a homologous series. The kinetic observations are related to the experimental temperatures by the linearized Arrhenius relationship

$$y_{ij} = \{\ln A\}_j - \{E/R\}_j x_{ij} + \varepsilon_{ij}$$

or more simply by

$$\underline{y}_j = X\underline{\theta}_j + \underline{\varepsilon}_j$$

where the observation vector is

$$\underline{y}'_j = (\ln k_{1j}, \ln k_{2j}, \ldots, \ln k_{nj})$$

the parameter vector is $\underline{\theta}'_j = (\{\ln A\}_j, \{-E/R\}_j)$ and the design matrix is

$$X' = \begin{bmatrix} 1 & 1 & \cdots & 1 \\ 1/T_1 & 1/T_2 & \cdots & 1/T_n \end{bmatrix}$$

if all the m-data sets are taken at the same $1 \leq j \leq n$ temperatures. The enthalpy and entropy estimates are related to the θ-estimates by

$$\Delta S_j^{\neq} = R\{\ln A\}_j - R\ln(kTe/h) = R\theta_{1j} + C_1$$

$$\Delta H_j^{\neq} = E_j - RT = -R\theta_{2j} + C_2$$

which may be summarized by

$$\underline{\psi}_j = Z\underline{\theta}_j + \underline{C}$$

where the thermodynamic parameter vector is $\underline{\psi}_j' = (\Delta S_j^{\neq}, \Delta H_j^{\neq})$, the additive constant vector is $\underline{C}' = (-R\ln(kTe/h), -RT)$ and the matrix Z is

$$Z = \begin{bmatrix} R & O \\ O & -R \end{bmatrix}$$

Given that the rate constants are measurable to within an experimental error ε_j the usual linear regression problem is

$$\underline{y}_j = X\underline{\theta}_j + \underline{\varepsilon}_j \quad \text{where } \underline{\varepsilon}_j \sim \text{NID}(\underline{O}, \sigma_j^2)$$

with the least squares solution

$$\hat{\underline{\theta}}_j = (X'X)^{-1}X'\underline{y}_j$$

$$\hat{\underline{\psi}}_j = Z\hat{\underline{\theta}}_j + \underline{C}$$

for each member $1 \leq j \leq m$ of the homologous series.

Unfortunately, this is where the proper application of regression analysis usually ends and experimenters improperly try to fit the $(\Delta\hat{H}, \Delta\hat{S})$ estimates to a linear relationship. Since $\underline{\varepsilon}_j \sim \text{NID}(\underline{O}, \sigma_j^2)$ all information about $\underline{\psi}_j$ is contained in the estimated first and second moments. The first moment, the maximum likelihood estimator (MLE) or least squares estimate (LSE) $\underline{\psi}_j$, has been properly calculated in the past. Ignorance of the second moment has led to incorrect conclusions, however. The second moment can be illustrated graphically as a joint likelihood region $\ell(\underline{\psi}_j|\underline{y}_j,X)$ for the location of parameter values $\underline{\psi}_j$ given the data $\underline{y}_j$ and experimental design X or as a joint probability region $p(\hat{\underline{\psi}}_j|\underline{\psi}_j,\sigma_j^2,X)$ for the location of the estimates $\hat{\underline{\psi}}_j$ given the true value $\underline{\psi}_j$, the variance σ_j^2 and the choice of experimental settings X.

Such joint confidence regions are displayed in Figure 1a for a single member j = 1 of a homologous series. To understand how m = 7 (as plotted in Figure 1b) such least squares estimates would be distributed in the absence of a chemical effect (or equivalently if the variation of rate constants by measurement errors was much greater than the variation by a chemical effect) we consider the following analysis using a sampling theory approach. Since $\underline{y}_j = X\underline{\theta} + \underline{\varepsilon}$ and $\underline{\varepsilon} \sim NID(\underline{O},\sigma^2)$, then $\underline{y}_j \sim NID(X\underline{\theta},\sigma^2)$ setting $\sigma_j^2 = \sigma^2$ and $\underline{\theta}_j = \underline{\theta}$ for all j because we are assuming all data originate from the same source. The probability distribution of observations for the jth data set is

$$p_j(\underline{y}_j|\underline{\theta},\sigma,X) = \frac{1}{(\sqrt{2\pi}\ \sigma^2)^n} \exp\left\{-\frac{1}{2\sigma^2}(\underline{y}_j-X\underline{\theta})'(\underline{y}_j-X\underline{\theta})\right\}$$

Since $\hat{\underline{\theta}}_j$ is a linear combination of $\underline{y}_j$, $\hat{\underline{\theta}}_j = (X'X)^{-1}X'\underline{y}_j$, then $\hat{\underline{\theta}}_j \sim NID(\underline{\theta},(X'X)^{-1}\sigma^2)$ and the probability distribution of the θ-estimates is

$$p_j(\hat{\underline{\theta}}_j|\underline{\theta},\sigma,X) = \frac{|X'X|^{1/2}}{2\pi\sigma^2} \exp\left\{-\frac{1}{2\sigma^2}(\hat{\underline{\theta}}_j-\underline{\theta})'X'X(\hat{\underline{\theta}}_j-\underline{\theta})\right\}$$

Finally, $\hat{\underline{\psi}}_j$ is a linear combination of $\hat{\underline{\theta}}_j$, $\hat{\underline{\psi}}_j = Z\hat{\underline{\theta}}_j + \underline{C}$, so that $\hat{\underline{\psi}}_j \sim NID(\underline{\psi},Z(X'X)^{-1}Z\sigma^2)$ yielding

$$p_j(\hat{\underline{\psi}}_j|\underline{\psi},\sigma^2,X) = \frac{|X'X|^{1/2}}{2\pi\sigma^2} \exp\left\{-\frac{1}{2\sigma^2}(\hat{\underline{\psi}}_j-\underline{\psi})'Z^{-1}X'XZ^{-1}(\hat{\underline{\psi}}_j-\underline{\psi})\right\}$$

This distribution of m = 7 such estimates is given by

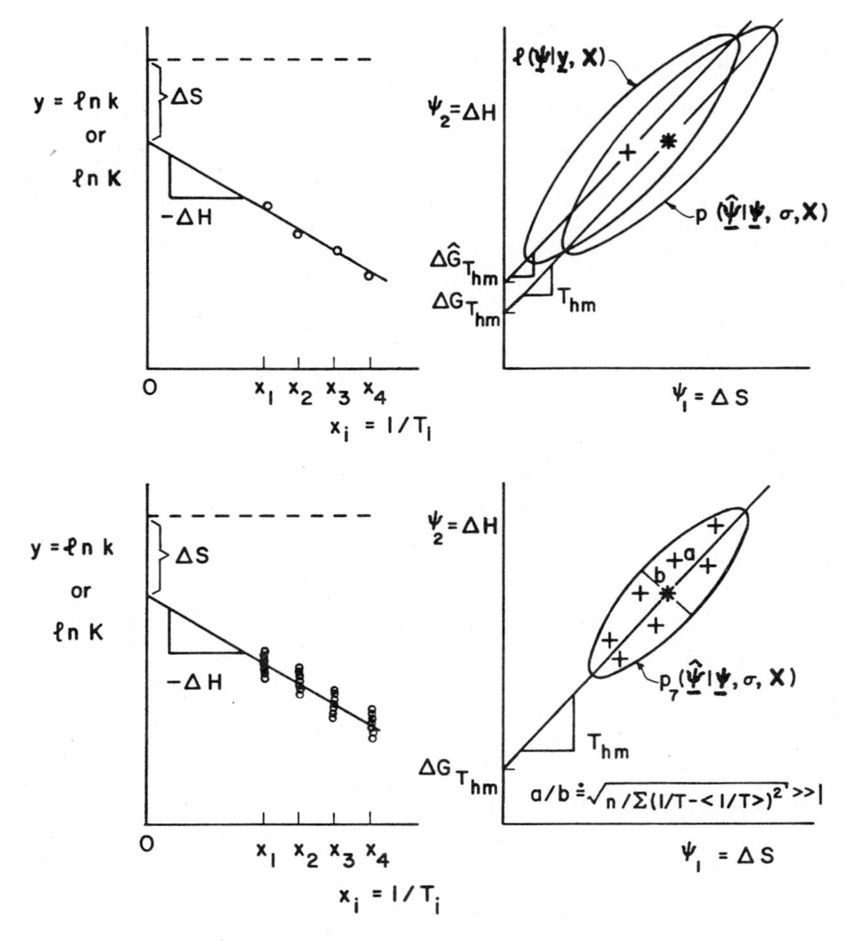

Figure 1. (a) The Arrhenius or van't Hoff regression problem in ln k *versus* 1/T *coordinates and the resultant joint likelihood and probability regions in the coordinates of the estimated parameters. (b) The Arrhenius or van't Hoff regression problem in* ln k *versus* 1/T *coordinates for the case of* $j = m = 7$ *replicates at each* $i = n = 4$ *temperature and the joint probability region for the seven parameter estimates (designated by +) given the true value (designated by *) in the coordinates of the estimated parameters. Notice that the ratio a/b of major to minor axes of each ellipse in (a) and (b) is identical and a function only of the choice of experimental temperatures.*

$$p_{j=m=7}(\hat{\underline{\psi}}|\underline{\psi},\sigma^2,X) = \prod_{j=1}^{m=7} p_j(\hat{\underline{\psi}}_{\underline{j}}|\underline{\psi},\sigma^2,X)$$

$$= \frac{|X'X|^{m/2}}{(2\pi\sigma^2)^m} \cdot \exp\left\{- \frac{m}{2\sigma^2} (\hat{\underline{\psi}}_{\underline{j}}-\underline{\psi})'Z^{-1}X'X\, Z^{-1}(\hat{\underline{\psi}}_{\underline{j}}-\underline{\psi})\right\}$$

This probability distribution contains all the information available about the thermodynamic parameter estimates $\hat{\underline{\psi}}' = (\underline{\Delta\hat{S}^{\neq}},\underline{\Delta\hat{H}^{\neq}})$ for a homologous chemical series for which the variation of rate constants is due much more to measurement errors than to the effect of an extrathermodynamic relationship. We will now use this information to derive the statistical compensation equation and to show that such estimates are highly correlated for the usual experimental temperature ranges even in the absence of an extrathermodynamic effect.

<u>The Correlation Coefficient</u>. The correlation coefficient for a single data pair $(\Delta\hat{S}^{\neq}_j,\Delta\hat{H}^{\neq}_j)$ and for a complete j = m data set $(\underline{\Delta\hat{S}^{\neq}},\underline{\Delta\hat{H}^{\neq}})$ are determined from the elements of their respective variance-covariance matrices

$$V(\hat{\underline{\psi}}_{\underline{j}}) = V\begin{bmatrix}\Delta\hat{S}^{\neq}_j \\ \Delta\hat{H}^{\neq}_j\end{bmatrix} = Z(X'X)^{-1}Z\sigma^2 = \begin{bmatrix}\Sigma(1/T)^2 & \Sigma 1/T \\ \Sigma 1/T & n\end{bmatrix}\frac{R^2\sigma^2}{|X'X|}$$

$$V(\hat{\underline{\psi}}) = V\begin{bmatrix}\underline{\Delta\hat{S}^{\neq}} \\ \hline \underline{\Delta\hat{H}^{\neq}}\end{bmatrix} = Z(X'X)^{-1}Z\frac{\sigma^2}{m} = \begin{bmatrix}\Sigma(1/T) & \Sigma 1/T \\ \Sigma 1/T & n\end{bmatrix}\frac{R^2\sigma^2}{m|X'X|}$$

For either case, the correlation coefficient is the same. For the complete data set, the joint probability region has decreased in area because the variance has decreased from σ^2 to σ^2/m, however. The correlation coefficient between enthalpy and entropy estimates for any series for which the variation of rate constants by measurement errors being much greater than by the effect of an extrathermodynamic relationship is

$$\rho = \frac{\text{Cov}\,(\underline{\Delta \hat{H}^{\neq}}, \underline{\Delta \hat{S}^{\neq}})}{\sqrt{V(\underline{\Delta \hat{H}^{\neq}})\,V(\underline{\Delta \hat{S}^{\neq}})}} = \frac{\Sigma 1/T}{\sqrt{n\Sigma (1/T)^2}}$$

where the estimated correlation coefficient

$$r = \frac{\Sigma(\Delta\hat{H}_j - \langle\Delta\hat{H}\rangle)(\Delta\hat{S}_j - \langle\Delta\hat{S}\rangle)}{\sqrt{\Sigma(\Delta\hat{H}_j - \langle\Delta\hat{H}\rangle)^2\,\Sigma(\Delta\hat{S}_j - \langle\Delta\hat{S}\rangle)^2}}$$

is an estimate of the population parameter ρ.

$$\lim_{m\to\infty} r = \rho$$

Thus, if a correlation coefficient r is estimated from $(\underline{\Delta\hat{H}}, \underline{\Delta\hat{S}})$ data pairs such that a confidence interval for r includes the value of ρ, the linear distribution of enthalpy-entropy estimates is probably due to the propagation of measurement errors and not due to any detectable extrathermodynamic effect.

That ρ should be near unity is illustrated in Figure 2 for data taken on the oximation of methyl thymyl ketone (30). A measurement error in one rate constant upon replication would result in a slightly different slope estimate and the intercept estimate at $1/T = 0$ would change correspondingly. The very high correlation, $\rho = 0.99991$, between the slope estimate and the intercept estimate is a consequence of the fact that the data are taken far from the origin over a very narrow temperature range on an absolute scale.

The Statistical Compensation Equation. As shown in Figure 1, the shape and orientation of $\ell_j(\underline{\psi_j}|\underline{y_j}, X)$, $p_j(\underline{\hat{\psi}_j}|\underline{\psi}, \sigma, X)$ and $p_7(\underline{\hat{\psi}}|\underline{\psi}, \sigma, X)$ are all identical, merely the size and location change from one to the other. In particular, the shapes of the joint confidence regions are elliptical because the fitted model is linear and the orientations are a function of the experimental design, which is the choice of experimental temperatures at which the rate constants were measured.

The ratio of major to minor axes of these elliptic regions is determined by a consideration of the ratio of eigenvalues of $Z^{-1}X'XZ^{-1}$ such that $\lambda_1 > \lambda_2$ and is found to a good approximation to be

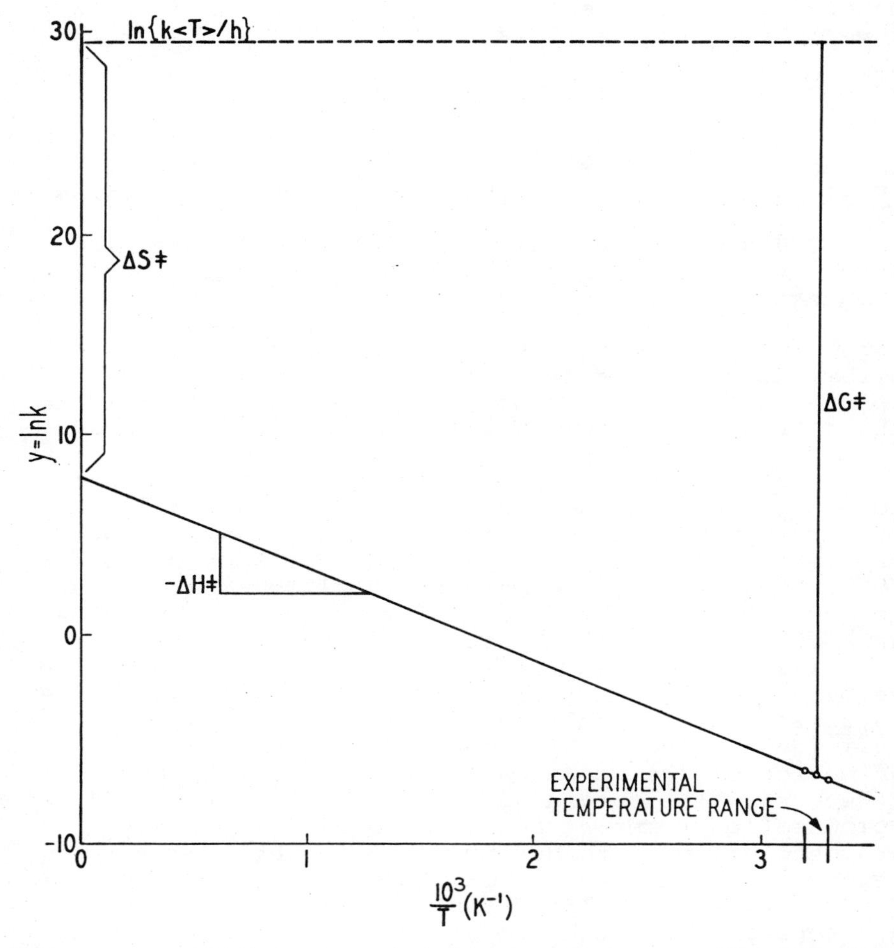

Figure 2. Geometric interpretation of the parameter estimates. The indicated lengths are proportional to $\Delta G^{=}$ and $\Delta S^{=}$ and the indicated slope is proportional to $-\Delta H^{=}$. These data for the oximation of methyl thymyl ketone (30) indicate a strong dependence of the intercept estimate on the slope estimate because the data were taken over a very small temperature range far from the origin. The three data points are designated by dots. Reprinted with copyright permission by Nature (25) and the Journal of Physical Chemistry (26).

$$a/b = \sqrt{\lambda_1/\lambda_2} \simeq \sqrt{n/\Sigma(1/T-<1/T>)^2}$$

For the usual experimental temperature ranges of organic chemistry this ratio is usually of the order of 10^4. Hence, the joint probability regions appear as line segments and are well characterized by the line that describes the major axis of the ellipse. A canonical analysis of the dispersion matrix $Z^{-1}X'XZ^{-1}$ reveals that this line is

$$\underline{\Delta H} = T_{hm}\underline{\Delta S} + \Delta G_{T_{hm}}$$

and differs from the extrathermodynamic equation, $\underline{\Delta H} = \beta\underline{\Delta S} + \Delta G_\beta$, only in the value of the slope parameter. Hence, of the estimated compensation temperature β is near the harmonic mean of the experimental temperatures T_{hm} the compensation that is detected might only be the statistical compensation between parameter estimates that occurs because the range of the independent variable was too small to validate the extrathermodynamic model in this parameter space.

Separation of the Chemical from the Statistical Compensation. Because any extrathermodynamic effect is strongly confounded with the statistical compensation effect in the enthalpy-entropy parameter space for the usual ranges of experimental temperatures used in organic chemistry, biochemistry, and even heterogeneous catalysis, some statisticians have attempted to solve the problem for the value of the compensation temperature in the original ln k versus 1/T space (19, 20,22,24). The resulting normal equations yield unwieldy nonlinear solutions that are better for the detection of the presence of an extrathermodynamic effect than for obtaining good numerical values of a compensation temperature. Others (15,31-34) have proposed criteria to determine if an observed compensation is of chemical origin or is just the statistical artifact. We find that the two compensations are separable through a translation of the intercept and that the compensation temperature and its confidence interval can be solved for exactly using likelihood theory and the chemical Maxwell equations.

The problem is to choose an intercept for which the slope and intercept estimates are not correlated. The intercept at the arithmetic mean of the independent variable has this property. Thus, we rewrite the linearized Arrhenius equation in the form

$$\ln k_{ij} = \{\ln A - E/RT_{hm}\}_j - \{E/R\}_j(1/T_i - \langle 1/T\rangle) + \varepsilon_{ij}$$

where the independent variable is now $(1/T_i - \langle 1/T\rangle)$.

The slope is still a measure of the enthalpy, but the intercept is now a measure of the free energy at the harmonic mean of the experimental temperatures.

$$\Delta G^{\neq}_{T_{hm}} = -RT_{hm}\{\ln A - E/RT_{hm}\} + RT_{hm}\ln(kT_{hm}e/h) - RT_{hm}$$

The model is now

$$\underline{\eta}_j = X\underline{\theta}_j = W\underline{\zeta}_j, \quad \underline{Y}_j = \underline{\eta}_j + \underline{\varepsilon}_j$$

where the parameter vector is $\underline{\zeta}'_j = (\{\ln A - E/RT_{hm}\}_j, \{-E/R\}_j)$ and the design matrix is

$$W' = \begin{bmatrix} 1 & 1 & \cdots & 1 \\ 1/T_1 - \langle 1/T\rangle & 1/T_2 - \langle 1/T\rangle & \cdots & 1/T_n - \langle 1/T\rangle \end{bmatrix}$$

The slope and intercept parameters are related to the thermodynamic parameters by

$$\underline{\Psi}_j = A\underline{\zeta}_j + \underline{B}$$

where the thermodynamic parameter vector is $\underline{\Psi}'_j = (\Delta G^{\neq}_{T_{hm_j}}, \Delta H^{\neq}_j)$, the additive constant vector is

$\underline{B}' = (RT_{hm}\ln(kT_{hm}e/h) - RT_{hm}, -RT)$ and

$$A = \begin{bmatrix} -RT_{hm} & 0 \\ 0 & -R \end{bmatrix}$$

Proceeding as before, we determine that there is no correlation between $\underline{\Delta G^{\neq}_{T_{hm}}}$ and $\underline{\Delta H^{\neq}}$ estimates after a consideration of the variance-covariance matrix

$$V(\underline{\hat{\Psi}}) = \begin{bmatrix} T_{hm}^2\Sigma(1/T - \langle 1/T\rangle)^2 & 0 \\ 0 & n \end{bmatrix} \frac{R^2\sigma^2}{m|W'W|}$$

$$\rho = \frac{\mathrm{Cov}(\underline{\Delta\hat{G}^{\neq}_{T_{hm}}}, \underline{\Delta H^{\neq}})}{\sqrt{V(\underline{\Delta\hat{G}^{\neq}_{T_{hm}}})V(\underline{\Delta\hat{H}^{\neq}})}} = \frac{0}{T_{hm}\sqrt{n\Sigma(1/T-\langle 1/T\rangle)^2}} = 0$$

The ratio of variances between these estimates is found to be a constant that depends only on the choice of experimental temperatures

$$\lambda_j = \frac{\sigma^2_{\Delta\hat{H}_j}}{\sigma^2_{\Delta\hat{G}_{T_{hm_j}}}} = \frac{(\Sigma 1/T)^2}{n\Sigma(1/T-\langle 1/T\rangle)^2}$$

such that if the same experimental temperatures are chosen for all experiments, all such estimate pairs will have the same ratio of variances (i.e., $\lambda_j = \lambda$ for all j if $T_{ij} = T_i$ for all j). Since the Maxwell equations are linear relationships between the thermodynamic potentials H, G, E, and A and the properties S, T, P, and V, an extrathermodynamic linear relationship between any two must also be reflected by a linear extrathermodynamic relationship between any other two. In particular, if an extrathermodynamic relationship

$$\underline{\Delta H} = \beta\underline{\Delta S} + \Delta G_\beta$$

exists, then by substitution with the Gibbs equation,

$$\underline{\Delta H} = \gamma\underline{\Delta G} + (1-\gamma)\Delta G_\beta$$

where the diagnostic parameter γ is related to the compensation temperature β by

$$\gamma = 1/(1-T/\beta)$$

and

$$\Delta G_\beta = \underline{\Delta H} - \beta\underline{\Delta S} = \underline{\Delta H} + \gamma T_{hm}\underline{\Delta S}/(1-\gamma)$$

Thus, if $\lambda \neq 1$, the Gibbs equation is insufficient to explain detected chemical behavior and an extra-thermodynamic effect if detected. No statistical compensation exists between $\underline{\Delta\hat{G}_{T_{hm}}}$ and $\underline{\Delta\hat{H}}$ where $\underline{\Delta\hat{H}}$ may be evaluated at any temperature, including $T = T_{hm}$. To test the null hypothesis, H_o: $\gamma = 1$, $\underline{\Delta\hat{H}}$ must be regressed on $\underline{\Delta\hat{G}_{T_{hm}}}$ to estimate the slope γ and intercept $(1-\gamma)\Delta G_\beta$. Least squares is an incorrect procedure, because there is uncertainty in both variables. The errors are uncorrelated, however, and the ratio of variances is known, see Figure 3a. The likelihood function is maximized in this case by

$$\min_{a,\gamma} \sum_{j=1}^{m} \frac{(\Delta\hat{H}_j - a - \gamma\Delta\hat{G}_{T_{hm_j}})^2}{s^2_{\Delta\hat{G}_{T_{hm_j}}}(\lambda+\gamma^2)}$$

This type of problem was first solved for the scope estimate by Lindley (35) and later commented on by others (36-38). To obtain a confidence interval for γ (and hence β) the distribution of γ must be determined. Creasy (39) solved this type of problem in transformed coordinates, which correspond in our case to ΔH versus $\sqrt{\lambda}\Delta G_{T_{hm}}$ (see Figure 3b), in which the joint probability regions are circular, that is, the errors propagate randomly with no preferential direction. From the distribution of the correlation coefficient, the distribution of the angle ϕ that a regression line would make through such a plane is determined. The distribution of the slope γ is determined from the relationship between γ and ϕ. For our case, we extend this line of reasoning one more step and from the relationship between γ and β, the distribution and hence the maximum likelihood value and confidence interval of the compensation temperature β is determined.

The Regression Algorithm. Given kinetic or equilibrium data, ($\underline{k}$,$\underline{T}$) or ($\underline{K}$,$\underline{T}$), the following algorithm may be used to obtain maximum likelihood estimates and their $(1-\alpha)$ confidence intervals for ϕ,

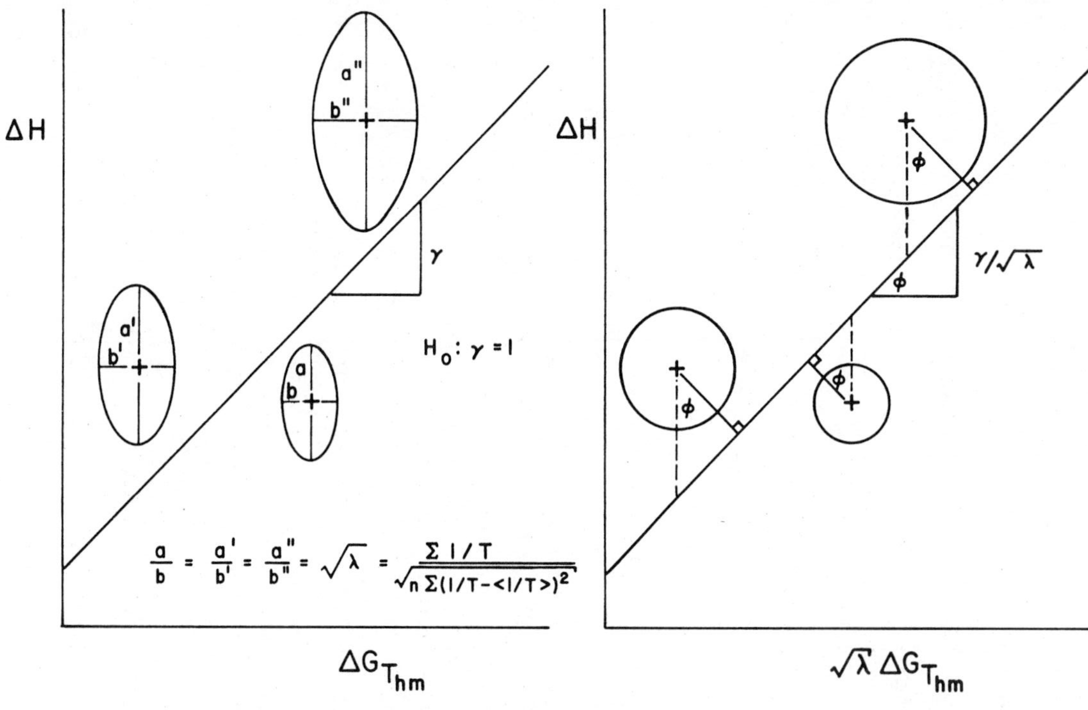

Figure 3. *(a) The linear regression problem in ΔH-$\Delta G_{T_{hm}}$ coordinates is one for which joint confidence regions $l(\hat{\Psi}|y,X)$ have a constant ratio λ of major to minor axes when the data are sampled at identical temperatures. (b) In ΔH-$\sqrt{\lambda}\,\Delta G_{T_{hm}}$ coordinates the maximum likelihood fit to a line is found by minimizing the sum of squares of residuals which are the perpendiculars to the regression line. Creasy (39) solved for the distribution of the slope estimate from the distribution of the correlation coefficient in similar coordinates.*

γ, a, ΔG_β and the compensation temperature β for a homologous series of chemical data with $1 \leq j \leq n$ temperatures.

1. Regress $y_{ij} = \ln k_{ij}$ or $\ln K_{ij}$ onto $(1/T_i - \langle 1/T \rangle)$ to obtain parameter estimates $\underline{\hat{\zeta}}_j$ and residual sum of squares s_j^2

$$\underline{\hat{\zeta}}'_j = \left[\frac{\sum_{i=1}^{n} y_{ij}}{n} \; , \; \frac{\sum_{i=1}^{n} y_{ij}(1/T_i - \langle 1/T \rangle)}{\sum_{i=1}^{n} (1/T_i - \langle 1/T \rangle)^2} \right]$$

and

$$s_j^2 = \frac{\sum_{i=1}^{n} (y_{ij} - \hat{\zeta}_{1j} - \hat{\zeta}_{2j}\{1/T_i - \langle 1/T \rangle\})^2}{(n-2)}$$

$$T_{hm} = n/\Sigma 1/T_i = \langle 1/T \rangle^{-1}$$

2. Then calculate enthalpy and free energy estimates from the slope and intercept estimates. For kinetic data

$$\Delta \hat{G}^{\neq}_{T_{hm_j}} = -RT_{hm}\hat{\zeta}_{1j} + \{RT_{hm}\ln(kT_{hm}e/h) - RT_{hm}\}$$

$$\Delta \hat{H}^{\neq}_j = -R\hat{\zeta}_{2j} - RT$$

and for equilibrium data

$$\Delta \hat{G}^{\circ}_{T_{hm_j}} = -RT_{hm}\hat{\zeta}_{1j}$$

$$\Delta \hat{H}^{\circ}_j = -R\hat{\zeta}_{2j}$$

3. The data may be plotted with joint confidence regions determined by the elliptic equation

$$(\underline{\Psi}_j - \underline{\hat{\Psi}}_j)' A^{-1} W' W A^{-1} (\underline{\Psi}_j - \underline{\hat{\Psi}}_j) = 2s_j^2 F(2, n-2, 1-\alpha)$$

or the data may be plotted along with standard deviation increments from the maximum likelihood estimates from Step 2.

$$s_{\Delta\hat{G}_{hm_j}} = \sqrt{V(\Delta\hat{G}_{T_{hm_j}})} = (T_{hm}R)\sqrt{s_j^2/n}$$

$$s_{\Delta\hat{H}_j} = \sqrt{V(\Delta\hat{H}_j)} = R\sqrt{s_j^2/\Sigma(1/T-\langle 1/T\rangle)^2}$$

If a linear regression appears to be justified from this plot, then proceed with 4 and 5. If a nonlinear functionality is to be fitted, use an appropriate weighted nonlinear technique. The weighting factors are $s^2_{\Delta\hat{H}_j}=V(\Delta\hat{H}_j)$ from above.

4. Calculate the following to find maximum likelihood estimates using Lindley's solution (35).

$$\lambda = (\Sigma 1/T)^2/(n\Sigma(1/T-\langle 1/T\rangle)^2)$$

$$s_{GG} = \Sigma\Delta G_j^2/s_j^2 - (\Sigma\Delta G_j/s_j^2)^2/\Sigma 1/s_j^2$$

$$s_{HH} = \Sigma\Delta H_j^2/s_j^2 - (\Sigma\Delta H_j/s_j^2)^2/\Sigma 1/s_j^2$$

$$s_{HG} = \Sigma\Delta H_j\Delta G_j/s_j^2 - \Sigma\Delta H_j/s_j^2\Sigma\Delta G_j/s_j^2/\Sigma 1/s_j^2$$

$$\Theta = (s_{HH}-\lambda s_{GG})/2s_{HG}$$

$$\hat{\gamma} = \Theta \pm \sqrt{\Theta^2+\lambda}\ \mathrm{sgn}(\sqrt{\Theta^2+\lambda'}) = \mathrm{sgn}(s_{HG})$$

$$\hat{\phi} = \tan^{-1}(\hat{\gamma}/\sqrt{\gamma'})$$

$$\hat{a} = (\Sigma\Delta H_j/s_j^2-\hat{\gamma}\Sigma\Delta G_j/s_j^2)/\Sigma 1/s_j^2$$

$$\Delta\hat{G}_\beta = \hat{a}/(1-\hat{\gamma})$$

$$\hat{\beta} = T_{hm}/(1-1/\hat{\gamma})$$

5. Finally (1-α)100% confidence intervals may be calculated from the following upper and lower bound estimates using Creasy's solution (39).

$$\hat{\phi}_{\substack{U\\L}} = \hat{\phi} \pm \frac{1}{2}\sin^{-1}\left[2t_{\alpha/2,m-2}\sqrt{\frac{\lambda(s_{HH}s_{GG}-s_{HG})}{(m-2)\{(\lambda s_{GG}-s_{HH})^2 + s_{HG}^2\}}}\right]$$

$$\hat{\gamma}_{\substack{U\\L}} = \sqrt{\lambda}\tan\hat{\phi}_{\substack{U\\L}}$$

$$\hat{a}_{\substack{U\\L}} = (\Sigma\Delta H_j/s_j^2 - \hat{\gamma}_{\substack{L\\U}}\Sigma\Delta G_j/s_j^2)/\Sigma 1/s_j^2$$

$$\Delta\hat{G}_{\beta_{\substack{U\\L}}} = \hat{a}_{\substack{U\\L}}/(1-\hat{\gamma}_{\substack{L\\U}})$$

$$\hat{\beta}_{\substack{U\\L}} = T_{hm}/(1-1/\hat{\gamma}_{\substack{L\\U}})$$

This regression algorithm gives maximum likelihood estimates and their confidence intervals even though there is error in both variables, because an additional restraint is placed on the system--the ratio of variances of dependent to independent variables is a known constant, a function of the experimental temperatures. This restraint holds so long as each system is sampled at identical temperatures. If this ratio λ becomes very large, the estimates will converge on the weighted least squares estimates.

An interesting sidelight is the minimum likelihood estimate, the "worst" value of a parameter given the data. This estimate is given by (35,36)

$$\hat{\gamma}^* = \Theta \pm \sqrt{\Theta^2+\lambda} \quad \mathrm{sgn}(\sqrt{\Theta^2+\lambda}) = -\mathrm{sgn}(s_{HG})$$

$$\beta^* = T_{hm}/(1-1/\gamma^*)$$

Because of the high correlation between enthalpy-entropy estimates, the application of least squares to these enthalpy-entropy estimates will yield numerical values of the compensation temperature that are nearer the minimum likelihood value rather than the maximum likelihood value. Thus by misuse of least squares, a valuable statistical technique, the "worst" numerical value of a chemical parameter has usually been reported in the literature rather than the "best" numerical value.

Our analysis of 37 reported enthalpy-entropy compensations revealed that only three had compensation temperatures significantly different than the harmonic mean of the experimental temperatures by an analysis in the ΔH-ΔS plane (26) and only 7 had detectable chemical compensations by an analysis in the ΔH-$\Delta G_{T_{hm}}$ plane (27).

Application to Chemical Examples

To illustrate the necessity of the proper regression procedure and proper correlation analysis, we compare a data set that clearly has a linear chemical compensation with one that clearly does not show such an extrathermodynamic effect. The validity of such an effect for this second example has been debated many times in the literature (7,9,16,18).

We find that data for the hydrolysis of ethyl benzoate (1) display a linear extrathermodynamic effect but data for the hydrolysis of alkyl thymyl ketones (30) do not. The results of a comparative correlation analysis are listed in Table I. As expected from our previous arguments on the correlation coefficient, both data sets display significant correlations r in ΔH-ΔS coordinates that approximate the expected correlation coefficient ρ due to the propagation of errors. Only the hydrolysis data has a significant (40) estimated correlation coefficient in ΔH-$\Delta G_{T_{hm}}$ coordinates, however. This finding indicates that the observed enthalpy-enthropy correlation for the oximation data is a result of only the propagation of measurement errors.

Table I. Correlation Coefficients*

Reaction	$\Delta\hat{H}^{\neq}$-$\Delta\hat{S}^{\neq}$ ρ	$\Delta\hat{H}^{\neq}$-$\Delta\hat{S}^{\neq}$ r	$\Delta\hat{H}^{\neq}$-$\Delta\hat{G}^{\neq}_{T_{hm}}$ ρ	$\Delta\hat{H}^{\neq}$-$\Delta\hat{G}^{\neq}_{T_{hm}}$ r	m
1. Oximation of alkyl thymyl ketones (30)	0.9999	0.9724	0	-0.2273	7
2. Same as (1) but deleting the methylated compound	0.9999	0.9988	0	0.0770	6
3. Hydrolysis of ethyl benzoate (1)	0.9988	0.9987	0	0.9929	12

*Reprinted with copyright permission by the Journal of Physical Chemistry (27).

The joint confidence regions, $\ell(\underline{\psi}|\underline{y},X)$ and $\ell(\underline{\Psi}|\underline{y},X)$, for the oximation and hydrolysis data are plotted in Figures 4 and 5 for comparison. The oximation data have much greater uncertainty than do the hydrolysis data. It is this greater uncertainty that is largely responsible for the apparent compensation in ΔH-ΔS coordinates. In fact the ratio of major to minor axes for the oximation data in ΔH-ΔS coordinates is a/b = 23252 causing the joint confidence regions to be well represented by the lines of their major axes.

The compensation temperature and other parameter estimates are compared in Table II for estimation by (a) the regression algorithm presented here, (b) weighted least squares of $\underline{\Delta\hat{H}}$ on $\underline{\Delta\hat{G}_{T_{hm}}}$ using $\underline{s^2_{\Delta\hat{H}}}$ as the weighting factors and (c) least squares of $\underline{\Delta\hat{H}}$ on $\underline{\Delta\hat{S}}$. Because $\lambda >> 1$ for both examples, the application of weighted least squares in the ΔH-$\Delta G_{T_{hm}}$ plane (b) gave estimates close to the maximum likelihood values (a). Also for both examples the minimum likelihood value of the compensation temperature $\hat{\beta}^*$ is near the harmonic mean of the experimental temperatures T_{hm} as expected. For both examples the value of the compensation temperature as determined by least squares of $\underline{\Delta\hat{H}}$ on $\underline{\Delta\hat{S}}$ (c) was biased toward $\hat{\beta}^*$ as expected, but for the oximation example the value of the compensation temperature as estimated by (c) was numerically much closer to the minimum likelihood estimate β^* than to the maximum likelihood estimate $\hat{\beta}$. That the confidence interval for $\hat{\beta}$ should appear to exclude $\hat{\beta}$ when no chemical compensation is detected is illustrated in Figure 6. If the probability distribution for the diagnostic parameter γ overlaps unity (recall H_o: $\gamma = 1$, where $\gamma = 1$ for no linear extrathermodynamic effect) the probability density for β is thinly distributed over all possible numbers such that the confidence interval for γ traces to a confidence interval for β that starts at a finite value, extends to infinity, returns from minus infinity, and finally returns to a finite value. The MLE of β is then somewhere in that interval. When this happens H_o cannot be rejected and the probability density of β is distributed so thinly that the probability of detecting a compensation temperature in a reasonably finite interval is infinitesimal, and hence the probability that a linear extrathermodynamic effect is detected is essentially zero.

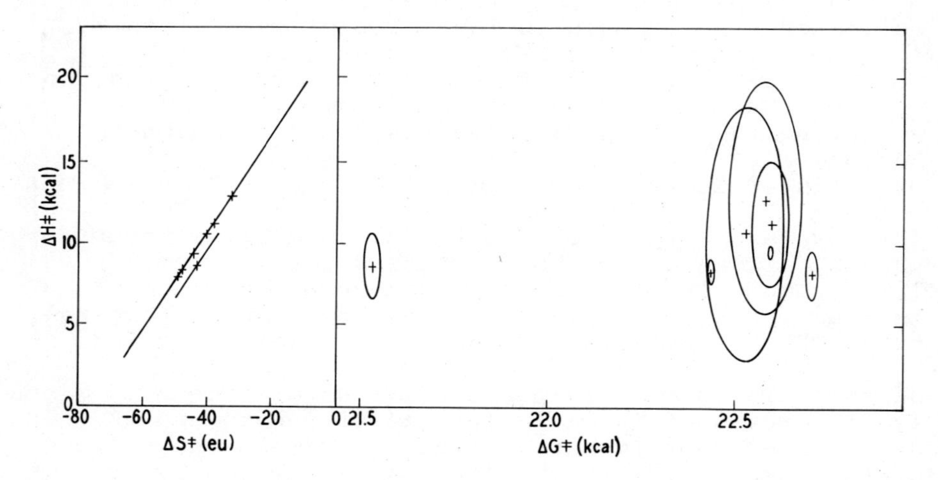

Nature
Journal of Physical Chemistry

Figure 4. The 50% joint confidence regions for the oximation of alkyl thymyl ketones (30). The $\Delta H^{=}$–$\Delta S^{=}$ ellipses are so narrow that they appear as lines. Departure of the methylated compound from a common $\Delta G^{=}$ value causes it to fall off the statistical compensation "line" between $\Delta H^{=}$ and $\Delta S^{=}$ estimates. Notice that $\Delta G^{=}$ is estimated more precisely than $\Delta H^{=}$. All values were calculated for $T = T_{hm} = 308.1$ K (25, 27).

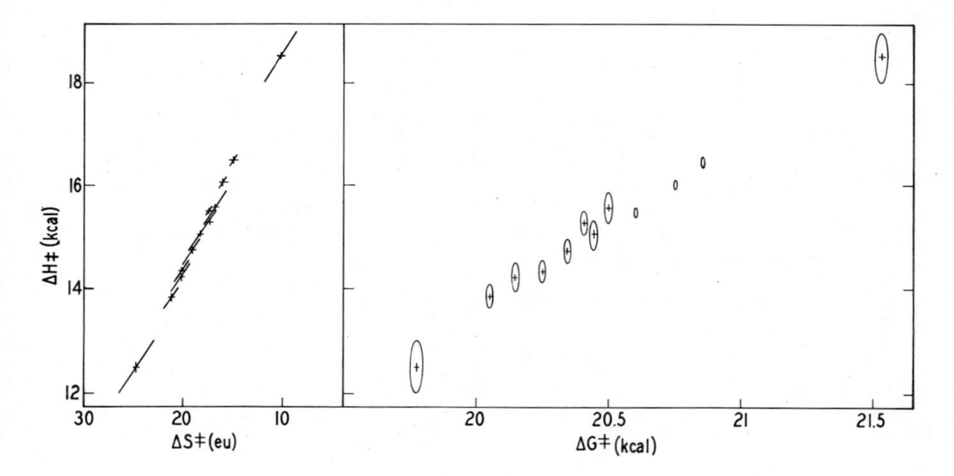

Journal of Physical Chemistry

Figure 5. Plots of thermodynamic parameter estimates and their respective 50% confidence regions for the hydrolysis of ethyl benzoate (1). The linear structure in the $\Delta H^{\neq}$–$\Delta G^{\neq}$ plot indicates that a linear chemical compensation is detected. The data are evaluated at $T = T_{hm} = 292.6$ K (27).

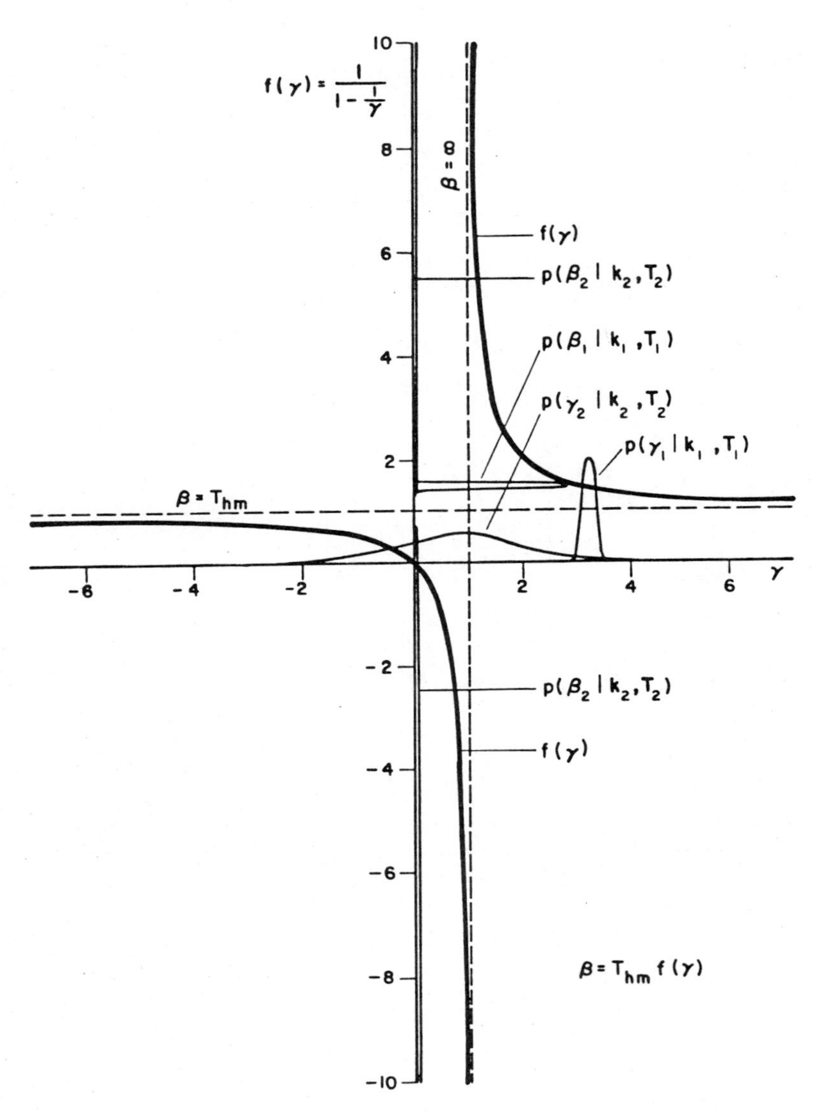

Figure 6. Probability density functions for γ and β. The bold line is the function $f(\gamma)$ through which the well-behaved probability density functions $p(\gamma|k,T)$ are mapped into either well-behaved or skewed density functions $p(\beta|k,T)$. The footnote "1" represents parameters obtained from the hydrolysis of ethyl benzoate (1), and "2" represents parameters obtained from the oximation of thymyl ketones (30). Reprinted with copyright permission by the Journal of Physical Chemistry (27).

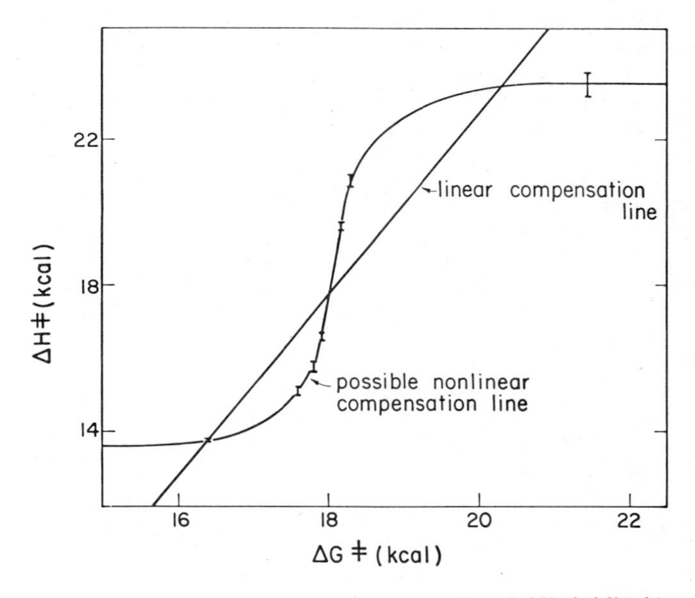

Figure 7. Carbinol formation from Malachite dyes (41). The data are plotted along with 50% confidence ellipses or one standard deviation increments. The variances of the $\Delta G^{\ddagger}_{T_{hm}}$ estimates are too small to be noticed on this plot. A nonlinear functionality appears to be suggested by the data. The data are evaluated at $T = T_{hm} = 297.7$ K. Reprinted with copyright permission by the Journal of Physical Chemistry (27).

Finally, we show in Figure 7 that nonlinear relationships are easily visualized as well when the thermodynamic data are plotted in the ΔH-$\Delta G_{T_{hm}}$ plane. If a preferred nonlinear function is to be fit to such data, weighted least squares may be used to obtain numerical parameter estimates when $\lambda >> 1$, which is the usual case. Such structured nonlinear functions are virtually impossible to distinguish from random (unstructured) scatter in ΔH-ΔS coordinates because of the dominant statistical compensation between parameter estimates in those coordinates.

Conclusions

The history of enthalpy-entropy compensation is one characterized by a strong misuse of fundamental statistical tools (the method of least squares and correlation analysis) to reach incorrect conclusions about the detection and validity of observed compensations between enthalpy and entropy estimates. Any chemical effect in the enthalpy-entropy plane is strongly confounded with the linear statistical compensation pattern due to the limited experimental temperature ranges suitable for most chemical investigations. Enthalpy and entropy estimates are still frequently plotted versus one another to display false correlations (compensations) for organic, biochemical, and heterogeneous catalytic reactions, however.

Because both the relevant chemical identities and the choice of experimental temperatures are available to a scientist prior to data analysis, he can and should choose to perform his correlation analysis and regression in the ΔH-$\Delta G_{T_{hm}}$ plane to detect the presence of both linear and nonlinear relationships between any thermodynamic variables for both kinetic and equilibrium data.

Literature Cited

1. Fairclough, R.A., Hinshelwood, C. N., J. Chem. Soc. (1937) 538.
2. Fairclough, R. A., Hinshelwood, C. N., J. Chem. Soc. (1937) 1537.
3. Raine, H. C., Hinshelwood, C. N., J. Chem. Soc. (1939) 1378.
4. Boudart, M., "Kinetics of Chemical Processes," pp 179-182, 194-198, Prentice-Hall, Englewood Cliffs, N.J., 1968.
5. Hammett, L. P., "Physical Organic Chemistry," 2nd Ed., pp 391-408, McGraw-Hill, New York, 1970.
6. Laidler, K. J., "Chemical Kinetics," 2nd Ed., pp 251-253, McGraw-Hill, New York, 1965.
7. Leffler, J. E., Grunwald, E., "Rates and Equilibria of Organic Reactions," pp 315-402, Wiley, New York, 1963.
8. Thomas, J. M. and Thomas, W. J., "Introduction to the Principles of Heterogeneous Catalysis," pp 263-265, 386,413, Academic Press, New York, 1967.

9. Leffler, J. E., J. Org. Chem. (1955) 20, 1202
10. Blackadder, D. A., Hinshelwood, C., J. Chem. Soc. (1958) 2720.
11. Blackadder, D. A., Hinshelwood, C., J. Chem. Soc. (1958) 2728.
12. Petersen, R. C. Markgraf, J. H., Ross, D. S., J. Amer. Chem. Soc. (1961) 83, 3819.
13. Brown, R. F., Newsom, H.C., J. Org. Chem. (1962) 27, 3010.
14. Brown, R. F., J. Org. Chem. (1962) 27, 3015.
15. Petersen, R. C., J. Org. Chem. (1964) 3133.
16. Leffler, J. E., Nature (1965) 205, 1101.
17. Exner, O., Nature (1964) 201, 488.
18. Exner, O., Coll. Czech. Chem. Comm. (1964) 29, 1094.
19. Exner, O., Nature (1970) 227, 366.
20. Exner, O., Coll. Czech. Chem. Comm. (1972) 37, 1425.
21. Wold, S., Chem. Scr. (1972) 2, 145.
22. Wold, S., Exner, O., Chem. Scr. (1973), 3, 5.
23. Wold, S., Chem. Scr. (1974) 5, 97.
24. Exner, O., Coll. Czech. Chem. Comm. (1975) 40, 2762.
25. Krug, R. R., Hunter, W. G., Grieger, R. A., Nature (1976) 261, 566.
26. Krug, R. R., Hunter, W. G., Grieger, R. A., J. Phys. Chem. (1976) 80, 2335.
27. Krug, R. R., Hunter, W. G., Grieger, R. A., J. Phys. Chem. (1976) 80, 2341.
28. Laidler, K. J., Trans. Faraday Soc. (1959) 55, 1725.
29. Ritchie, C. D., Sager, W., Prog. Phys. Org. Chem. (1964) 2, 323.
30. Craft, M. J., Lester, C. T., J. Amer. Chem. Soc. (1951) 73, 1127.
31. Garn, P. D., J. Therm. Anal.(1975) 7, 475.
32. Good, W. and Ingham, D. B. Electrochem. Acta (1975), 20, 57.
33. Gorbacher, V. M., J. Therm. Anal. (1975) 8, 585.
34. Gorbacher, V. M., Izv. Sib. Otd. Akad. Nauk. SSSR, Ser. Khim. Nauk. (1975) 5, 164.
35. Lindley, D. V., J. Roy. Stat. Soc. Suppl. (1947) 9, 218
36. Madansky, A., J. Amer. Stat. Assn. (1959) 54, 1973.
37. Cochran, W. G., Technometrics (1968) 10, 637.
38. Davies, O. L., Goldsmith, P. L., "Statistical Methods in Research and Production," pp 208-210, Hafner, New York, 1972.
39. Creasy, M. A., J. Roy. Stat. Soc. B (1956) 18, 65.

40. David, F. N., "Tables of the Correlation Coefficient," Cambridge University Press, Cambridge, England, 1954.
41. Idlis, G. S., Ginzburg, O. F., Reakc. Sposobnost Org. Sojedin. (1965) 2, 54.

11

How to Avoid Lying with Statistics

ALLAN E. AMES and GEZA SZONYI

Polaroid Corp., 750 Main St., Cambridge, MA 02139

I. Introduction

Statistical analysis of randomly varying data has become commonplace in the age of calculators and computers. Such analyses are often carried out routinely using built-in calculator functions or standard computer subroutines. Parameters derived in such computations usually include the average (mean) and the standard deviation, characteristic of the central tendency and the variability of individual data sets. To compare two data sets with each other, the differences between their averages and the ratio of their variances are used customarily. Little thought is usually given to the fact that the computation of the above parameters presupposes that the data analyzed are essentially normally distributed and that this distribution is monomodal, i.e., showing essentially one major peak or central value only. If this is not the case, and standard (parametric) methods are applied for the evaluation of such data, the results obtained will represent an incorrect picture. In other words, the evaluator is: "lying with statistics", usually without being aware he is doing so (1). To make matters worse, nearly all built-in programs for calculators, as well as the majority of the subroutines and programs for computers are of the parametric type. Nowhere in the instruction manuals is the fact adequately stressed that the usage of parametric methods presupposes known, usually normal, distribution which must be ascertained before these methods are applicable.

This paper will show that for many applications the use of distribution-free (nonparametric) methods may be preferable and easier to use than the conventional parametric approach (1). (Throughout this paper, the term "parametric tests" pertains to the t-Test and the F-test, used with the appropriate tables (2, 3, 4, 5, 6, 7, 8, 9)). We will also provide a simple method for testing the normality of the distribution of data sets. Examples will be given to demonstrate the penalties which can be incurred when parametric methods are used for the analysis of nonparametric data.

II. Methodology

Normally distributed data are often represented graphically by the well-known, bell-shaped Gaussian curves (10, 11, 12, 13, 14). To obtain such a curve, the data in question are usually sorted in ascending order and then grouped into classes. The number of items in each class, i.e., the class frequencies, are then plotted against class midpoints in such standard frequency plots. Another method of representation for the same data is the use of cumulative frequency plots. Cumulative frequencies are obtained by adding ("cumulating") for each class all the class frequencies up to that point, divided by the total number of data in the data set (10, 15), as shown in Figure 1.

This paper deals with two applications of the cumulative distribution. In one case, a known continuous distribution--the normal distribution--is being compared to an unknown discrete distribution in order to determine its normality or deviation from it. In another case, two unknown cumulative discrete distributions are being compared to each other to determine their sameness or difference. For both cases the same test statistic is applicable (16, 17, 18):

$$T = \sup |F(x) - S(x)|$$

where: F(x) is the cumulative distribution function of an unknown distribution; S(x) is the cumulative distribution function of either a known (e.g. normal) or an unknown distribution; and, sup = supremum

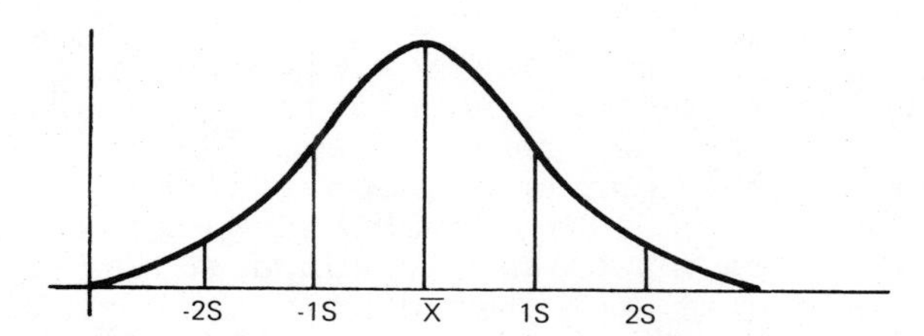

Figure 1a. Representation in terms of standard frequencies

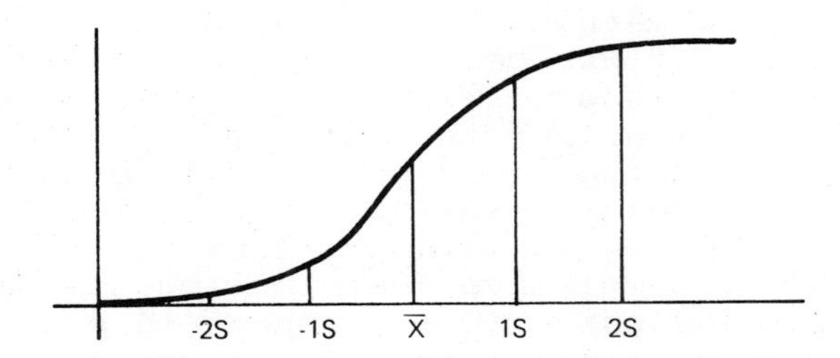

Figure 1b. Representation in terms of cumulative frequencies

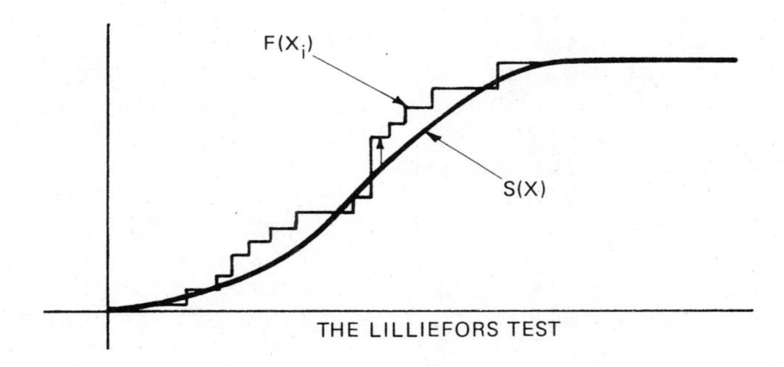

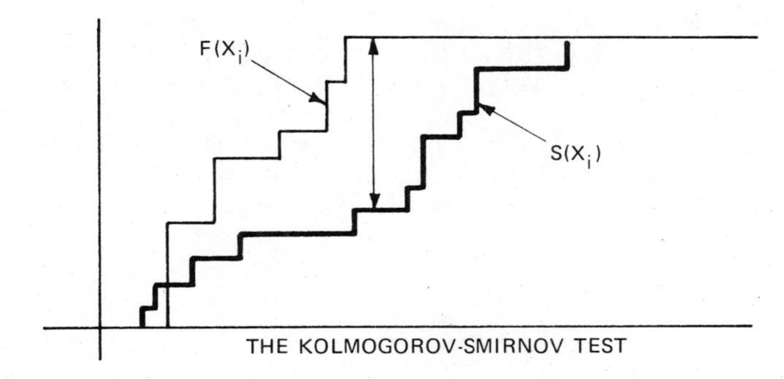

Figure 2. Graphical interpretation of Smirnov-type test statistics

signifies maximum, and together with the two vertical lines symbolizes that one should take the maximum difference enclosed by these lines. Thus, T is the maximum vertical distance between the two cumulative distribution graphs. This computed T, or a similarly derived value, is subsequently compared to appropriate tabulated values at selected confidence levels. The two distributions are regarded as being different if this T is greater than the corresponding tabulated value, otherwise, the opposite holds true.

A number of methods have been described in the literature to test whether a given data set can be considered to be essentially normally distributed or not. Table 1 summarizes the most important of these methods. As seen from this table, all methods are based on the use of cumulative distribution functions, except the Shapiro-Wilk method. (The Chi Square test uses the normal cumulative distribution to compute theoretical frequencies (19).) There is no overall agreement, based on the literature surveyed, as to which method is "best" for all possible empirical distributions (20, 21, 22). Comparison of the various normality tests has lead to the result that in some cases, the same data are considered normal by some tests and not normal by others (22). However, the Chi Square test is generally regarded to be inferior to all other tests (20, 21, 22).

The extent to which parametric methods can be used for borderline normality cases has not been described in the literature surveyed. This area needs substantial exploration and, at this stage, informed intuition is as good a guide as any.

This paper deals with a simple test for normality, the Lilliefors Test (23, 24), and not with the question of what normality test to select. We have found this test satisfactory to help us to decide when to use and when not to use parametric statistics in data analysis. To analyze nonnormally distributed data, we have been using the Kolmogorov-Smirnov Two Sample Test (16, 17, 18), which compares two data sets of unknown distribution for their sameness or difference using the cumulative distribution approach. The principle of these two tests is presented graphically in Figure 2.

The Lilliefors Test of Normality.

The computational procedure for the Lilliefors Test involves the following steps (16, 17, 18):

a) Arrange the data set to be tested in ascending order.

b) Calculate the following quantities:

The mean:

$$\bar{X} = \frac{1}{N} \Sigma x_i = \frac{1}{N} \Sigma x_i f_i$$

The standard deviation:

$$S = \frac{\Sigma (x_i - \bar{x})^2}{N - 1} = \frac{\Sigma x_i^2 - (\Sigma x_i)^2/N}{N - 1}$$

$$= \frac{\Sigma (x_i - \bar{x})^2 f_i}{N - 1} = \frac{\Sigma x_i^2 f_i - (\Sigma x_i f_i)^2/N}{N - 1}$$

The Z-Statistics for each distinct data value (Z(I)):

$$Z_i = \frac{x_i - \bar{x}}{S}$$

where:

x_i = individual datum, or class midpoint of data arranged by class frequencies.

f_i = frequency of datum or data class

N = total number of data in the data set

c) for each datum (X(I)), or data class within the data set, compute its cumulative frequency (CUM) with the formula:

$$\text{CUM(I)} = \frac{i - 1}{N}$$

where: i = the rank order of the datum, or its frequency, or the frequency of the data class.

This means that the first entry of the data set is omitted, but cumulative frequencies are computed for all other data values, i.e., from: (i - 1) to N. The cumulative frequencies thus generated represent the unknown distribution function: $F(x_i)$.

d) Using the Z_i-values computed for every datum or data class, determine the standardized cumulative normal distribution value (XNORM) corresponding to it. These standardized normal values can be obtained from tables of the Normal Probability Functions (25), using the IBM Scientific Subroutined NDTR (26), or writing your own subroutine utilizing the error function (27). These values represent the cumulative normal distribution function: $S(x_i)$.

e) Compute the absolute difference (DIF) between CUM and XNORM, and determine the largest value (BIG) among these values. This represents: T, the test statistics for the Lilliefors Test.

f) By using the Lilliefors Table (23, 24, 28), determine whether this computed T is statistically significant or not, at the selected confidence level. If T is not significant, the data analyzed can be regarded as being essentially normally distributed so that both parametric and nonparametric methods can probably be used for the analysis of the data set in question. If T is significant only nonparametric methods should be used.

A simple illustration of this procedure is presented in Table 2, using previously published data (22).

The procedure described is applicable to all types of data sets, i.e , either sets of individual data, or sets of data specified through frequencies within defined classes.

Steps a), b) and d) of this procedure are common to all normality tests given in Table 1, with the exception of the Chi Square and the Shapiro-Wilk methods, i.e., to all Empirical Distribution Function (EDF) statistics which use cumulative frequencies (22). Having done these basic calculations, all EDF based tests can be carried out easily with a computer or calculator.

Table 1

Principal Methods for Normality Testing

Test Method	Computing Formula	Symbols Used	Remarks	References
Chi Square	$\chi^2 = \sum \frac{(O_j - T_j)^2}{T_j}$	O_j=Observed class frequency T_j=Theoretical class frequency	The data are arranged in descending order and then subdivided into arbitrary classes, each containing at least 5 numbers. The "observed frequencies" are the number of data in each class. The "theoretical frequencies" are those based on the cumulative normal distribution function. The computed χ^2 is evaluated using a χ^2 -Table at k-3 degrees of freedom, where: k=number of classes. This method requires a relatively large sample size (at least 15-30) and is considered a test of low power.	19,23,24
Lilliefors	$\max D_i^- = S(x_i) - \frac{i-1}{n}$	i=sequential data points or frequencies (i-1)/n=computes the cumulative distribution values of the data set $S(x_i)$=corresponding normal cumulative distribution values	The data are arranged in ascending order and the D_j^- values computed by the formula given, i.e, not considering the first data cell. The largest D_j^- value obtained is evaluated using the Lilliefors Table. This method can be used with data sets given in terms of frequencies or individual data. Sample sizes as low as 4 can be tested. This method will fail if the first data cell contains a disproportionately large number of data compared to the rest of the set.	23,24

Table 1

(continued)

Test Method	Computing Formula	Symbols Used	Remarks	References
Kolmogorov D^+ (Kolmogorov One-Sample Statistics)	$\max D_i^+ = S(x_i) - \frac{i}{n}$	Symbols as given and: i/n=computes the cumulative distribution of the data set	The data are arranged in ascending order and the D_i^+-values computed, as given by the formula. The largest D_i^+-value obtained is evaluated using the Kolmogorov Table (two-sided). The method is usable with sample sizes as low as 2, with data given individually or as frequency distributions. This method can supplement the Lilliefors Test if it is suspected that it may fail. A modified version of this test multiplies the largest D_i^+-value by a correction factor and evaluates this modified value using a special table (22,46).	22,46,47,48 49,50
Kolmogorov D (Modified Kolmogorov, E_n)	$D_i = \max (D^+, D^-)$	Symbols as given	The largest D_i^+ or D_i^- value, calculated by the above formulas is chosen. This value is multiplied by a correction factor and evaluated using a special table (22,46). Usable for individual data or those given as frequency distributions.	22,46
Kuiper	$V = D^+ + D^-$	Symbols as given	The sum of the largest D_i^+ and D_i^- is calculated by the formulas given. This sum is multiplied by a correction factor and evaluated using a special table (22,46). Usable for individual data or those given as frequency distributions.	22, 46
Cramer-von Mises (Cramer-Smirnov)	$W^2 = \frac{1}{12_n} + \sum \left[S(x_i) - \frac{2_i - 1}{n} \right]^2$	Symbols as given and: $(2_i-1)/n$=computes the cumulative distribution values of the data set.	The data are arranged in ascending order. The W^2 computed is evaluated using appropriate tables. From this W^2, a modified W^2 can be calculated, and this is evaluated using a special table (22,46). With the standard table, sample sizes as low as 2 can be evaluated. Usable for individual data or those given as frequency distributions.	22,46,47,48 49,50,51

Table 1

(continued)

Test Method	Computing Formula	Symbols Used	Remarks	References
Watson	$U^2 = W^2 - n(\bar{S}(x) - 1/2)^2$	Symbols as given, and: $\bar{S}(x) - S(x_i)/n$	This is a modification of the Cramer-von Mises method. The computed U^2 is modified and this modified U^2 is then evaluated using a special table (22,46). Usable for individual data or those given as frequency distribution.	22,46
Anderson-Darling	$A^2 = -\{\Sigma(2_i - 1)[\ln S(x_i) + \ln(1 - S(n+1-i))]\}/n - n$	Symbols as given	The data are arranged in ascending order. The A^2 computed is modified and this modified value is evaluated using a special table (22,46). Usable for individual data or those given as frequency distributions.	22,46
Shapiro-Wilk	For even number of data: $n=2K$ $W = \frac{\left[\sum_1^k a_{n-i+1}(x_{n-i+1} - x_i)\right]^2}{(x_i - \bar{x})^2}$ For odd number of data: $n=2K+1$ $W = \frac{[a_n(x_n - x_1) + \ldots + a_{k+2}(x_{k+2} - x_k)]^2}{\Sigma(x - \bar{x})^2}$	x_i=data; $\bar{x} = x_i/n$; a_n=special coefficients	The data are arranged in ascending order. The special coefficients needed for the calculation and the table needed to evaluate the W computed are only accessible via the reference listed. Samples as low as 3 can be evaluated.	32,33,34

NOTE: The various tests for normality have been recently reviewed by Shapiro et al. (32, 33, 34), Schuster (35), Stephens (22), Dyer (20), Shapiro and Francis (21), Klimko and Antle (36) and Govindarajulu (37).

Table 2

ILLUSTRATION OF THE APPLICATION OF THE LILLIEFORS TEST

I	X(J)	Z(J)	CUM	XNORM	DIF
1	148	-0.9618	---	---	---
2	154	-0.7214	0.0909	0.2358	0.1448
3	158	-0.5611	0.1818	0.2877	0.10588
4	160	0.4809	0.2727	0.3156	0.0429
5	161	-0.4408	0.3636	0.3300	0.0336
6	162	-0.4008	0.4545	0.3446	0.1100
7	166	-0.2405	0.5454	0.4052	0.1402
8	170	-0.0802	0.6363	0.4681	0.1683
9	182	0.4008	0.7272	0.6554	0.0719
10	195	0.9218	0.8181	0.8212	0.0030
11	236	2.6451	0.9090	0.9960	0.0869

BIG = 0.1683

$\bar{X}$ = 172.

s = 24.95195

Lilliefors Table Values

80 = 0.206
85 = 0.217
90 = 0.230
95 = 0.249
99 = 0.284

Since BIG is smaller than all the tabulated values, it is regarded as being statistically not significant and the data set to be essentially normally distributed.

The Kolmogorov-Smirnov Test--Nonparametric Testing of Two Distributions for Differences or Similarity.

The Kolmogorov-Smirnov Two-Sample Test, also called "The Two-Sample Smirnov Test", is used when two data sets are being compared in order to determine whether the two sets belong to the same population or not regardless of their underlying distributions (16, 17, 18). This is the same as asking whether two data sets are in some way different from each other. To this end, cumulative distributions are calculated for each set, as well as the differences between the cumulative values. The largest absolute difference thus obtained is then compared to appropriate tabulated values at the desired confidence level. If this largest difference is statistically significant, the two sets are considered different, otherwise they are the same.

The computational procedure for the Two-Sample Smirnov Test is the following (16, 17, 18):

a) Sort both data sets in ascending order. The two sets are designated as vectors: X(I) and Y(J). where:

$$X(I) = X_1, X_2, X_3, \ldots, X_n$$

$$Y(J) = Y_1, Y_2, Y_3, \ldots, Y_m$$

and: n and m are the lengths of the vectors, i.e., the number of data in X and Y, respectively.

b) Form two new vectors: XX and YY, both of length: n + m. The XX vector is the augmented X vector containing zeros wherever there is a number in the ordered Y vector. Similarly, the YY vector is the augmented Y vector containing zeros wherever there is a number in the ordered X vector.

c) Compute the cumulative frequencies for both sets by

$$\frac{i}{n} \quad i = 1,n$$

and

$$\frac{i}{m} \quad j = 1,m$$

for each X and Y value. These cumulative frequencies are inserted into the cumulative distribution vectors X(I)C and Y(I)C at points corresponding to the location of data entries in the augmented X and Y vectors. X(I)C and Y(J)C are representative of the two unknown distribution functions:

$F(x_i)$ and $S(x_i)$

d) The difference (DIF) between X(I)C and Y(J)C is computed for each value of the augmented vectors. The largest positive or negative value is determined. This represents: T, the test statistic.

e) By using Tables of the Two-Sample Smirnov Test statistic (29, 30) determine whether this computed T is statistically significant or not at the confidence level selected. If T is not significant, the two data sets are regarded as being essentially belonging to the same population, otherwise they are not. (Different tables should be used for data sets of equal and unequal length.)

A simple illustration of this procedure is presented in Table 3, using previously published data (31).

Although the main application of the Smirnov Test is its use of comparing two nonnormally distributed data, it can be used to evaluate normally distributed data sets as well. When used on normally distributed data, differences between the two sets will be detected, regardless of whether these arise from differences in the means or differences in the variances. Furthermore, the Smirnov Test may detect differences between normally distributed data sets when parametric methods do not allow clear-cut decisions.

Table 3

ILLUSTRATION OF THE APPLICATION OF THE TWO-SAMPLE SMIRNOV TEST

N	I	X(I)	M	J	Y(J)	XX	X(I)C	YY	Y(J)C	DIF
9	1	7.6	15	1	5.2	0	0	5.2	0.067	-0.067
	2	8.4		2	5.7	0	0	5.7	0.133	-0.133
	3	8.6		3	5.9	0	0	5.9	0.200	-0.200
	4	8.7		4	6.5	0	0	6.5	0.267	-0.267
	5	9.3		5	6.8	0	0	6.8	0.3333	-0.3333
	6	9.9		6	8.2	7.6	0.1111	0	0.3333	-0.2222
	7	10.1		7	9.1	0	0.1111	8.2	0.4000	-0.2889
	8	10.6		8	9.8	8.4	0.2222	0	0.4000	-0.1778
	9	11.2		9	10.8	8.6	0.3333	0	0.4000	-0.067
	10			10	11.3	8.7	0.4444	0	0.4000	0.04444
				11	11.5	0	0.4444	9.1	0.467	-0.0226
				12	12.3	9.3	0.5555	0	0.467	0.0885
				13	12.5	0	0.5555	9.8	0.5337	0.0218
				14	13.4	9.9	0.6666	0	0.5337	0.1329
				15	14.6	10.1	0.7777	0	0.5337	0.244
						10.6	0.8888	C	0.5337	0.3551
						0	0.8888	10.8	0.6004	0.2884
						11.2	0.9999	0	0.6004	0.3995
						0	0.9999	11.3	0.6671	0.3328
						0	0.9999	11.5	0.7338	0.2661
						0	0.9999	12.3	0.8005	0.1994
						0	0.9999	12.5	0.8672	0.1327
						0	0.9999	13 4	0.9339	0.066
						0	0.9999	14.6	1.000	0

Smallest Difference: DIF = -0.3333 at Y(5)-6.8

Largest Difference: DIF = 0.3995 at X(9)=11.2

Two Sample Smirnov Test Table Values		
	90	0.4889
	95	0.5333
	99	0.64444

Based on the tabulated values, the two data set belong essentially to the same population.

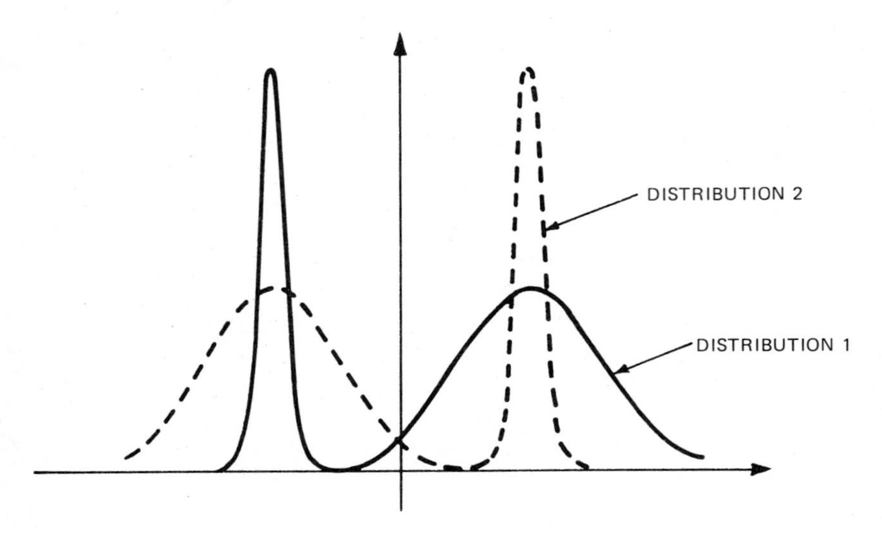

Figure 3. Two bi-modal distributions

III. Examples of How to Avoid Lying with Statistics

Let us now give some examples to demonstrate what happens when parametric statistics are used incorrectly.

A Monte Carlo Simulation was used to generate the two distributions shown graphically in Figure 3. The corresponding data sets and the results of the parametric t-Test and F-Test (8, 9) are shown in Table 4. These tests lead to the surprising conclusion that the two distributions are indistinguishable. If, however, the Two-Sample Smirnov Test is applied to the same data, the two data sets clearly differ from one another as one would expect from Figure 3. The two-sided test statistic is 0.5 which is significant at the 99th percentile. This contradiction between the results of the parametric and nonparametric tests indicates nonnormality of one or both data sets. Indeed, when the Lilliefors test is performed on these data, as shown in Tables 5 and 6, both sets show nonnormality. This means that only nonparametric methods are useful for comparing the data sets in question. To reassert the point made at the beginning

Table 4

PARAMETRIC EVALUATION OF TWO DATA SETS GENERATED BY MONTE CARLO SIMULATION

	Data Set 1	Data Set 2
	-50.06828	-157.53665
	-50.05978	-105.65264
	-50.04611	-104.06236
	-50.03810	- 76.14665
	-50.02966	- 76.76301
	-50.02777	- 72.76301
	-49.99072	- 65.85975
	-49.94664	- 57.93762
	-49.94615	- 55.93762
	-49.85856	- 54.95785
	-23.75126	30.14344
	- 5.20842	49.94298
	19.62010	49.95922
	19.66091	49.97837
	25.08628	49.98032
	26.31711	49.98634
	35.13998	50.00938
	61.47794	50.02284
	66.56312	50.04142
	78.51408	50.07682
$\bar{X}$	- 9.82857	- 17.29912
s^2	2176.994635	4992.86817
s	46.658271	70.660232

Tabulated t-values:	
$t_{38,90}$	1.3042
$t_{38,95}$	1.6860
$t_{38,97.5}$	2.0244
$t_{38,99}$	2.4286

Computed t: 0.39450

Tabulated F-values:	
$F_{19,19,90}$	1.8226
$F_{19,19,95}$	2.1686
$F_{19,19,97.5}$	2.5270
$F_{19,19,99}$	3.0282

Computed F: 0.43602

Table 5

LILLIEFORS TEST OF DATA SET 1

Lilliefors Test

J	X(J)	Z(J)	XNORM	CUM	DIF
1	-50.06828	-0.86241	0.19423	0.00000	0.19423
2	-50.05978	-0.86223	0.19428	0.05000	0.14420
3	-50.04611	-0.86193	0.19436	0.10000	0.09436
4	-50.03810	-0.86176	0.19440	0.15000	0.04440
5	-50.02966	-0.86138	0.19445	0.20000	0.00554
6	-50.02777	-0.86154	0.19446	0.25000	0.05553
7	-49.99072	-0.86075	0.19468	0.30000	0.10531
8	-49.94664	-0.85980	0.19494	0.35000	0.15503
9	-49.94615	-0.85979	0.19495	0.40000	0.20504
10	-49.85856	-0.85791	0.19546	0.44999	0.25453
11	-23.75126	-0.29837	0.38270	0.50000	0.11729
12	-5.20842	0.09904	0.53944	0.55000	0.01055
13	19.62010	0.63117	0.73603	0.60000	0.13603
14	19.66091	0.63205	0.73632	0.64999	0.06632
15	25.08628	0.74833	0.77287	0.70000	0.07207
16	26.31711	0.77471	0.78074	0.75000	0.03074
17	35.13998	0.96380	0.83242	0.80000	0.03242
18	61.47794	1.52829	0.93678	0.85000	0.08678
19	66.56312	1.53728	0.94921	0.89999	0.04921
20	78.51408	1.89341	0.97084	0.95000	0.02084

LILLIEFORS TEST OF SET 1 IS 0.2545.

Lilliefors Table Values

80	0.160
85	0.166
90	0.174
95	0.190
99	0.251

Table 6

LILLIEFORS TEST OF DATA SET 2

Lilliefors Test

J	X(J)	Z(J)	XNORM	CUM	DIF
1	-157.53665	-1.98467	0.02359	0.00000	0.02359
2	-105.65264	-1.25039	0.10557	0.05000	0.05557
3	-104.06236	-1.22709	0.10974	0.10000	0.00974
4	-76.14665	-0.93282	0.20247	0.15000	0.05247
5	-76.01930	-0.83102	0.20298	0.20000	0.00298
6	-72.76301	-0.76493	0.21624	0.25000	0.03376
7	-65.85975	-0.08724	0.24596	0.30000	0.05405
8	-57.93762	-0.57512	0.28260	0.35000	0.06739
9	-55.18869	-0.53622	0.29590	0.40000	0.10409
10	-54.55785	-0.53295	0.29703	0.44999	0.15296
11	30.14344	0.67141	0.74902	0.50000	0.24902
12	49.94298	0.95162	0.02935	0.55000	0.27935
13	49.95922	0.95185	0.82941	0.60000	0.22941
14	49.97837	0.95212	0.82948	0.64999	0.17948
15	49.98032	0.95215	0.82949	0.70000	0.12949
16	49.98634	0.95223	0.82951	0.75000	0.07951
17	50.00938	0.95256	0.82959	0.80000	0.02959
18	50.02284	0.95275	0.82964	0.85000	0.02035
19	50.04142	0.95501	0.82970	0.89999	0.07029
20	50.07682	0.95352	0.82983	0.95000	0.12016

LILLIEFORS TEST OF SET 2 IS 0.2793

Lilliefors Table Values

80	0.160
85	0.166
90	0.174
95	0.190
99	0.231

of this paper: the application of parametric methods to nonnormally distributed data may lead to drawing the wrong conclusions.

Our interest in nonparametric methods was caused by some production problems. Here it was important to decide whether a certain additive did or did not improve the quality of a certain product. The data, together with the parametric tests performed is shown in Table 7. Based on these tests, the particular additive apparently did not improve product quality. Since physically this did not make too much sense, the same data were reevaluated by subjecting them to the Two-Sample Smirnov Test. The highest computed value thus obtained was 0.319, as opposed to the tabulated values of: 0.21496(80), 0.24466(90), 0.271531(95), 0.30406(98) and 0.32527(99). Based on this test, the two data sets are indeed different and the additive in question did have an effect. Subjecting the data sets to the Lilliefors Test showed that both sets were nonormally distributed, giving computed values of: 0.31814 and 0.30726, resp., as opposed to tabulated values of: 0.11384(90), 0.1253(95) and 0.14581(99). Thus, the parametric tests are not applicable. What is particularly striking in this case is the large sample size which was taken to insure greater confidence in the results obtained, but which only reinforced an incorrect conclusion.

This example illustrates some important points. These are:

1. When physical reality apparently contradicts the statistical results obtained, it is incumbent on the applied statistician to reevaluate both his data and his approach.

2. Although a large sample size is generally desirable, it is no insurance for obtaining correct answers.

The preceding examples demonstrate that in certain circumstances the parametric tests may fail to indicate significant differences between distributions. The next example demonstrates the opposite situation, namely, that misapplied parametric tests can indicate significant differences between two distributions when in fact there are no differences. The specific circumstances are rather general, but we have noticed this to be the case if the parent distribution is skewed and the sample size small. Table 8 shows the data and the results of parametric and nonparametric tests performed on such data. Based on the parametric tests, the two data sets differ with regards to variability. This is, however, not the case based on the Two-Sample Smirnov Test. The Lilliefors Test shows that one of the data sets is nonnormally distributed.

IV. Conclusions

The preceding examples have demonstrated the problems that can arise when parametric statistics are used without proper foundation, that is, the independent verification of sufficient normality for the parametric tests to work correctly. But more important for the common "test" vs. "control" situation is that the more widely applicable nonparametric methods are more powerful (as compared to individual parametric tests) than the parametric tests for deciding when the "test" differs from the "control". (Some examples of the nonparametric statistics into data analysis can be found in the following references: (38, 39, 40, 41, 42, 43, 44, 45).)

As a general approach to answering the question "Is the 'test' different from the 'control'?", we recommend the following strategy. 1) For the first step in the intercomparison apply the two sample Smirnov test. If this step leads to the conclusion that no difference exists, no further analysis is necessary. 2) If the first step establishes that a difference exists and it is necessary to resolve the difference in terms of distributional parameters, apply the Lilliefors (or other appropriate) test. 3) If the data are normally distributed use the parametric tests as necessary. If not, find the transformations which do make the data properly parametric. If these transformations cannot be found, the questions being asked of the data cannot be answered and new questions must be found.

The computer programs used in the simulations and data analysis are available on request. Also available is the "Comprehensive Statistical Screening Program",

Table 7

PRACTICAL DATA AND RESULTS OF THE PARAMETRIC TESTS

Data 1	Data 2
5.65000	5.25000
5.17000	5.68000
5.75000	5.25000
5.40000	5.61000
5.75000	5.25000
5.08000	5.89000
5.75000	5.25000
5.00000	5.79000
5.75000	5.25000
5.86000	5.62000
5.75000	5.25000
5.20000	5.72000
5.75000	5.25000
5.20000	5.67000
5.75000	5.25000
5.25000	5.84000
5.75000	5.25000
5.09000	5.69000
5.75000	5.25000
5.33000	5.58000
5.75000	5.25000
5.52000	5.66000
5.75000	5.25000
5.25000	5.82000
5.75000	5.25000
5.26000	5.85000

Table 7

(continued)

Data 1	Data 2
5.75000	5.25000
5.35000	5.77000
5.75000	5.25000
5.20000	5.25000
5.76000	5.60000
5.75000	5.25000
5.27000	5.72000
5.75000	5.25000
5.09000	5.66000
5.75000	5.25000
5.43000	5.70000
5.75000	5.25000
5.26000	5.76000
5.75000	5.25000
5.28000	6.00000
5.75000	5.25000
5.40000	5.66000
5.75000	5.25000
5.46000	5.75000
5.75000	5.25000
5.20000	5.61000
5.75000	5.25000
5.33000	5.50000

Mean 1	5.50659	Mean 2	5.48219
Variance 1	0.07181	Variance 2	0.06057
Stand. Dev. 1	0.26797	Stand. Dev. 2	0.24611
Coeff. Var. 1	4.86643	Coeff. Var. 2	4.46933

Tabulated t-values:			Tabulated F-values:		
	$t_{98,90}$	1.2902		$F_{49,49,90}$	1.4496
	$t_{98,95}$	1.6606		$F_{49,49,95}$	1.6123
	$t_{98,97.5}$	1.9845		$F_{49,49,97.5}$	1.7691
	$t_{98,99}$	2.365		$F_{49,49,99}$	1.9724

Calculated t-value 0.47423

Calculated F-ratio 1.18553

Table 8

ANALYSIS OF SKEWED DATA

	Data Set 1	Data Set 2	
	-0.85837	-0.90514	
	-0.79696	-0.84636	
	-0.78426	-0.83421	
	-0.78426	-0.83421	
	-0.61911	-0.79696	
	-0.61911	-0.79696	
	-0.57363	-0.78426	
	-0.54223	-0.75840	
	-0.38865	-0.66278	
	-0.37018	-0.66278	
	-0.37918	-0.63386	
	-0.35139	-0.50991	
	-0.33228	-0.40681	
	-0.23139	-0.14344	
	0.14258	-0.07251	
	0.57340	-0.04782	
	1.21056	0.05691	
	1.58031	0.26831	
	1.58031	0.41030	
	4.11062	0.71030	
$\bar{X}$	0.078786	-0.412353	
s^2	1.488776	0.235497	
s	1.220154	0.485281	
Lilliefors	0.2503	0.1759	
Tabulated Lilliefors Values	$T_{20,90}$	0.174	
	$T_{20,95}$	0.190	
	$T_{20,99}$	0.231	
Tabulated t values	$t_{38,90}$	1.3042	
	$t_{38,95}$	1.686	computed t:
	$t_{38,97.5}$	2.024	1.673
Tabulated F values	$F_{19,19,95}$	2.168	
	$F_{19,19,97.5}$	2.527	computed F:
	$F_{19,19,99}$	3.028	6.322
Tabulated Two Sided Smirnov Values	$T_{20,20,80}$	0.30	
	$T_{20,20,90}$	0.35	computed T:
	$T_{20,20,95}$	0.40	0.349

combining Lilliefors, Kolmogorov-Smirnov, and a number of standard parametric tests.

We would like to express our thanks to the Polaroid Corporation for the permission to publish this work and to Jean Frederiksen of Polaroid for many helpful suggestions and stimulating discussions in connection with this paper.

Literature Cited

1. Mason, R.L., "FORTRAN Programs for Non-Parametric Studies", Naval Underwater Systems Center, New London, Connecticut, AD-769 649. National Techn. Information Service, U.S. Dept. of Comm., Springfield, Va., 1973.
2. Dixon, W.J. and Massey, F.J., "Introduction to Statistical Analysis", 3rd ed., pp. 114-119, McGraw-Hill Book Co., New York, 1969.
3. Dunn, O.J. and Clark, V.A., "Applied Statistics: Analysis of Variance and Regression", pp. 50-53, John Wiley & Sons, New York, 1974.
4. Dixon and Massey, loc. cit., pp. 109-113.
5. Dunn and Clark, loc. cit., pp. 53-55.
6. Natrella, M.G., "Experimental Statistics", National Bureau of Standards Handbook 91, pp. 4-1 to 4-14, U.S. Government Printing Office, Washington, D.C., 1963.
7. Natrella, loc. cit. pp. 4-8 and 4-9.
8. Owen, D.B., "Handbook of Statistical Tables", pp. 27-30, Addison-Wesley, Reading, Mass., 1962.
9. Owen, loc. sit., pp. 63-87.
10. Davies, O.L. and Goldsmith, P., "Statistical Methods in Research and Production", 4th rev. ed., pp. 12-16, Hafner Publishing Co., New York, 1972.
11. Dixon and Massey, loc. cit., pp. 9-10.
12. Haseloff, O.W. and Hoffman, H.J., "Kleines Lehrbuch der Statistik", p. 82, Walter de Gruyter & Co., Berlin, 1968.
13. Maisel, L., "Probability, Statistics and Random Processes", pp. 56-59, Simon & Schuster, New York, 1971.
14. Mosteller, F., Rourke, R.E.K. and Thomas, G.B., Jr., "Probability with Statistical Applications," pp. 259-268. Addison-Wesley Publishing Co., Reading, Mass., 1970.
15. Bethea, R.J., Duran, B.S. and Boullion, T.L., "Statistical Methods for Engineers and Scientists", p. 43, Marcel Dekker, Inc., New York, 1975.
16. Conover, W.J., "Practical Nonparametric Statistics", pp. 293-326, John Wiley & Sons, New York, 1971.
17. Siegel, S., "Nonparametric Statistics for the Behavioral Sciences", pp. 127-136, McGraw-Hill Book Co., New York, 1956.
18. Hollander, M. and Wolfe, D.A., "Nonparametric Statistical Methods", pp. 219-228, John Wiley & Sons, New York, 1973.

19. Dixon and Massey, loc. cit., pp. 243-244.
20. Dyer, A., Biometrika (1974) 61, pp. 185-189.
21. Shapiro, S.S. and Francis, R.S., J. Am. Stat. Assoc. (1972) 67, pp. 215-216.
22. Stephens, M.A., J. Am. Stat. Assoc. (1974) 69, pp. 730-737.
23. Liffiefors, H.W., J. Am. Stat. Assoc. (1967) 62, pp. 399-402.
24. Conover, loc. cit. pp. 302-306.
25. Beyer, N.H., "Handbook of Tables for Probability and Statistics", 2nd ed., pp. 125-134, The Chemical Rubber Co. Cleveland, Ohio, 1968.
26. IBM Application Program, System/360, Scientific Subroutine Package, Version III, Programmer's Manual, Program Number 360A-CM-03X, 5th ed., p. 78, 1970.
27. Alger, P.L., "Mathematics for Science & Engineering", pp. 303-304, McGraw-Hill Book Co., New York, 1969.
28. Conover, loc. cit. p. 398.
29. Conover, loc. cit. pp. 399 and 400-401.
30. Massey, F.J., Jr., Ann. of Math. Statistics (1952) 23, pp. 435-441.
31. Conover, loc. cit. p. 311.
32. Shapiro, S.S. and Wilk, M.B., Biometrika (1965) 52, pp. 591-611.
33. Shapiro, S.S., Wilk, M.B. and Chen, J.H., J. Am. Stat. Assoc. (1968) 63, pp. 1343-1372.
34. D'Agostino, R.B., Biometrika (1971) 58, pp. 341-348.
35. Schuster, E.I., J. Am. Stat. Assoc. (1973) 68, pp. 713-715.
36. Klimko, L.A. and Antle, C.E., Comm. Statistics (1975) 4, pp. 1009-1019.
37. Govindarajulu, Z., Comm. Statistic Theory Meth. (1976) A5, pp. 429-453.
38. Reed, A.H., Clinical Chemistry (1971) 17, p. 275.
39. Arnoldi, C.C., Acta Chirug. Sci. (1972) 138, p. 25.
40. Scholler, R., Path. Biol. (1973) 21, p. 375.
41. Wu, G.T., Twomey, S.L. and Thiers, R.E., Clinical Chemistry (1975) 21, pp. 315-320.
42. Smith, W.B., Tex. J. Sci. (1971) 22, p. 252.
43. Vestal, C.K., Monthly Weather Rev. (1971) 99, p. 650.
44. Britton, P.W., J. Sanit. Eng. (1972) 98, p. 717.
45. Locks, M.O. and Pauler, G.L., Proc. Am. Reliability and Maintainability Symposium (1974) 7, pp. 226-228.
46. Stephens, M.A., J. Roy. Stat. Soc., Ser. B, (1970) 32, pp. 115-122.
47. Conover, loc. cit., p. 397.
48. Massey, F.J., Jr., J. Am. Stat. Assoc. (1951) 46, pp. 69-78.
49. Owen, loc. cit., pp. 413-425.
50. Siegel, loc. cit. pp. 47-51.
51. Stephens, M.A. and Maag, M.R., Biometrika (1968) 5, pp. 428-430.

12

SIMCA: A Method for Analyzing Chemical Data in Terms of Similarity and Analogy

SVANTE WOLD and MICHAEL SJÖSTRÖM*

Research Group for Chemometrics, Institute of Chemistry, Umeå University, S-901 87 Umeå, Sweden

1. Introduction

Chemists study the behavior of molecules and mixtures of different molecules. Henceforth these mixtures or molecules are called chemical systems or objects. As in all branches of science, the increased quantification of measurements makes mathematics increasingly important both for the analysis of the measured data and for the prediction of the behavior of yet unstudied systems.

Roughly, we can distinguish 2 extremes of mathematical models employed for these purposes. The first type, best suited for rather simple systems, we can call global hard models. There are two contrasts; global-local and hard-soft. The term global is used to indicate that the models, in principle, describe the behavior of the system "everywhere". The polarization hard-soft is used mainly to describe the amount of information contained in the model. Hard models contain much information since they describe the system in terms of fundamental quantities. In addition, the deviation between the hard model and the measured data must not be larger than the errors of measurement.

Developed mainly in physics, global hard models describe a system in terms of fundamental physical quantities such as mass, charge, energy and time. Examples of this type of model employed in chemistry are quantum mechanical models and kinetic models in

* 1976–77 on leave to: Laboratory for Chemometrics, Department of Chemistry, BG-10, Seattle, WA 98195, USA.

the form of systems of differential equations.

The hard global models, when properly applied, have great advantages both in their far-reaching predictions and their interpretation in terms of fundamental quantities. However, their use is limited to systems of moderate complexity, since the complexity of the models -- in terms of the number of involved parameters and equations -- grows faster than exponentially when the number of components in the modelled systems increase.

Since global models of the type used in physics for a long time were the only mathematical models available, these were applied also in the study of very complex chemical systems involving many atoms and molecules, such as organic compounds reacting with each other in a solvent. For some aspects of such studies, for instance the kinetics, the success of the mathematical analysis was fair, but for other aspects, notably the relationship between structure and reactivity, the results of the mathematical analysis usually have been disappointing, in particular with regard to the predictions resulting from the analysis.

It was soon found that in order to apply the global hard models to systems of the complexity of common chemical systems, either the models had to be simplified to a degree that their interpretation was made dubious or the chemical system studied had to be simplified so far that it lost much of its relevance to the chemical problem that originated the study.

These difficulties with global hard models have made chemists continue to analyze their experimental data in a qualitative way. This qualitative analysis is often made in terms of analogy and similarity. Consider, as an example, an organic chemist studying properties of esters. When he starts the study of a new ester, he does not expect this ester to display an identical behavior to earlier studied esters. Rather, he expects a similar behavior. What is meant by "similarity" is difficult to state precisely, but the intuitive meaning is rather easy to see in each individual case.

In the same way, much of chemical theory is based on the concepts of similarity and analogy -- see for instance Theobald (1) for an illuminating discussion. Thus, the periodic system orders the elements in columns within which we find elements with "similar" chemical properties. Organic molecules with the same functional group, say COOR, display analogous properties with regard to reactivity, stability, spectra and so on.

With the large masses of quantitative data being produced in today's chemical laboratories -- mass spectrometry, NMR, UV, IR, Atomic Absorption, Electrophoresis and Amino-acid analysis, etc. -- the need to quantify the concepts of similarity and analogy becomes urgent. It is fairly easy to perceive a qualitative similarity between systems when one or two measurements are made on each of them, but extremely difficult when the number of data observed on each system becomes larger.

As will be seen below, the quantification of similarity and analogy can be made in a straight forward manner using the two concepts of classes -- i.e. local models -- and similarity models -- i.e. soft models. Before we go into this however, let us finish this section by fitting the different types of mathematical models into a philosophical framework.

The contrasts between global and hard, and local and soft models should not be seen as dichotomies of mutually exclusive mathematical tools. Rather, we see in each chemical study one part that is best studied by means of global hard models and one part that is best studied by means of soft models. Usually there is also a region of overlap where either type of model can be employed (see Figure 1). Let us illustrate this with the example of ^{13}C NMR which we discuss below. Here one problem is to utilize the spectra to determine whether a given molecule has exo or endo configuration. This is a problem of rather large complexity where the fundamental knowledge is rather scarce about which factors determine the difference in spectra between exo and endo molecules. Hence this problem is best approached using soft models. Another problem is the interpretation of the emerging patterns of the spectra in the two classes of exo and endo molecules. This is a problem where we require answers in terms of fundamental concepts and we therefore try to use hard models for this interpretation, say correlations with charge densities calculated by some quantum mechanical method.

In this way hard and soft models form complements to each other, each type being useful for the purpose they are designed.

Since the use of hard models (2,3) is treated by numerous authors in the chemical literature, we will not dwell more on this area but will here instead outline the basis and use of soft models (of a special type) in chemical investigations. We urge the reader to remember the complementary nature of the different types of mathematical models. The present treatment

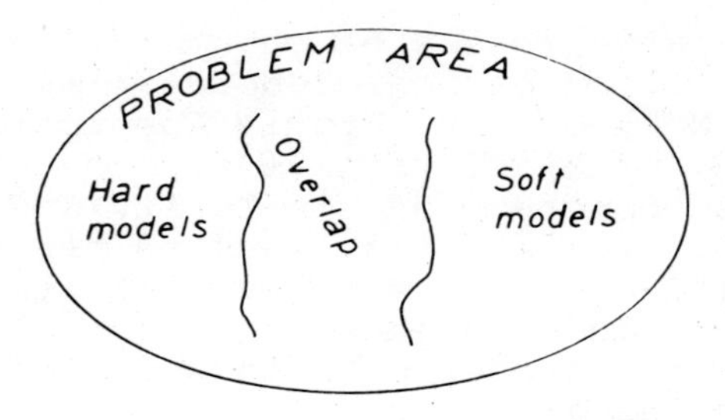

Figure 1. In a given problem different aspects are approached by means of different types of mathematical models

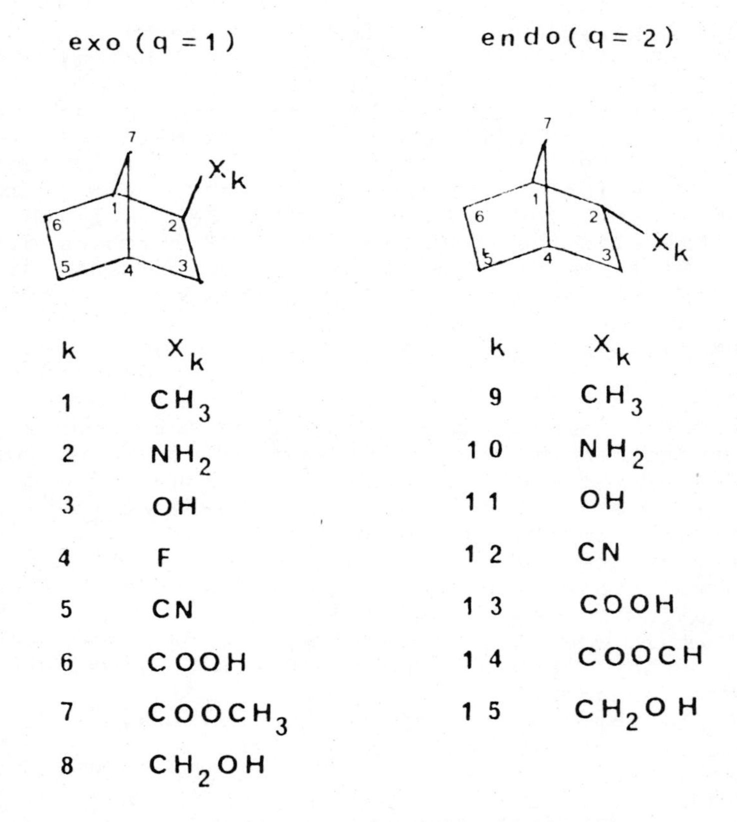

Figure 2. Learning set of 2-substituted norboranes. The exo *compounds form class 1 (q = 1) and the* endo *compounds class 2 (q= 2).*

thus discusses the use of mathematics in an area where this earlier has been difficult, we do not propose that the soft models should be used where hard models are appropriate.

However, there are many types of data analysis where, in our view, hard models are presently used rather inappropriately. This is because there is a lack of information about the existence and use of soft models. In such cases there might emerge an apparent conflict between the two types of models. This conflict will certainly be resolved when experience with the use of all kinds of mathematical models is common in all branches of chemistry.

2. Example

To illustrate the utility of soft models, we shall use the SIMCA-analysis of ^{13}C NMR data of norbornanes (Figure 2) (4). The data were analyzed to find out whether ^{13}C NMR data of these compounds could be used to determine if the structure of a particular compound is exo or endo and whether there existed consistent patterns for each type of molecules which could be utilized in the assignment of NMR-spectra. The data are shown in Table I. The reader is referred to section 5 for a more detailed discussion of which kind of information that one desires from this type of data analysis.

3. The Idea of Classes; Experimental Design and Similarity Models

The quantification of the similarity concept which we here propose (SIMCA) is based on three fundaments (5). The first is the experiment design which SIMCA presently can handle. Thus, we shall assume that on each object (chemical system) studied, a number of measurements have been made -- the same variables measured on all objects in the study. In the norbornane example, this number (henceforth denoted M) is seven; seven ^{13}C NMR shifts, one for each atom 1-7 in Figure 2, are collected for each of the studied compounds.

This design is rather common in chemical studies, the data collected in a study can often be arranged in a matrix as in Figure 3. As will be shown later in section 6.7 we need not assume that all positions in the matrix actually contain measured values, SIMCA works also with incomplete data.

Of critical importance for the success or failure

of the data analysis is that at least some of the variables have a relevance for the problem; i.e. that they are in some way related to the chemical problem under study. As will be seen below, one primary result of the data analysis is information about this relevance, whether the data contain any information at all and in such case, which parts that carry this information.

Second, we shall assume that the chemical systems have been ordered in classes in such a way that each class contain only similar systems (see Figure 3). Usually, there are also in the study a number of systems of uncertain class assignment. These we collect in the "test set" (see Figure 3) in order to later make an assignment by means of the mathematical models applied to the classes. This ordering into classes is usually easy in chemical investigations since chemists are accustomed to formulating their problems in terms of classes. In the norbornane example we have the two classes exo and endo compounds between which we wish to find the differences in ^{13}C NMR spectra.

Though we here illustrate the methodology by an example with two classes, the methods can also equally well be applied to problems with a single class or with a large number of classes. Depending on whether the number of classes is one or two or more, however, different types of information can be extracted from the data, a discussion of which is given in section 5.

It is evident that the concept of classes considerably simplifies the discussion, analysis and interpretation of the behavior of chemical as well as other natural systems. This is the reason for the popularity of this concept. Compounds are classified as organic or inorganic, aromatic or aliphatic, acids or bases, ionic or covalent and further into a large number of subclasses depending on the chemical problem studied. Reactions are classified according to "mechanisms", S_N1 or S_N2, solvent assisted or not, first or second order, acid or base catalyzed etc., etc. Thus, instead of considering a large number of individual cases or systems, one relates to a much smaller number of classes into one (or few) of which each case or system is assigned.

The third fundament is the existence of mathematical models which, under rather general assumptions, can approximate data observed on the similar objects in a single class. These similarity models, here called local and soft because they are limited to a single class and are of an approximate nature, can be derived by means of simple mathematical tools such as Taylor expansions. They can be considered as analogous to polynomials. In the same way as a polynomial of sufficiently high order always can approximate the variation of any continuous function in a limited interval, the similarity models shown in eq. (1) can, with sufficiently many terms, approximate any data matrix measured on a collection of similar systems, i.e., systems showing a limited diversity.

In conclusion, the similarity concept can presently be quantified in situations where a. measurements of the same type have been made on a number of systems, b. at least two measurements have been made on each system, c. the systems can be ordered into one or several classes plus a "test set" so that within each class there are only "similar" systems.

For each class we can then, on the basis of the data of the objects "known" to be in the class, construct a mathematical similarity model (see section 6.1). If we have Q classes in the study we thus obtain Q different models, each describing the data structure within a single class.

There are other ways to handle the classification problem, most of which are reviewed by Kanal (6) and Cacoullos (7). Some of these methods have been applied in chemical data analysis by Isenhour and Jurs (8), Kowalski (9,10) and others. The main difference between these methods and SIMCA which we here discuss is that the scope of SIMCA goes further than mere classification, namely to get an approximate description of the data structure within each class in terms of a quantitative model.

4. Graphical Representation

It is convenient to have a graphical representation of the data observed on the systems under study (Figure 3). The best way found so far is to represent the data measured on one system as a point in the M-dimensional space obtained when each variable is given one orthogonal coordinate axis. This space, the measurement space, is henceforth referred to as M-space. The ^{13}C NMR data in Table I would thus be represented as points in a 7-dimensional space; one point for each compound.

Though it is difficult to visualize spaces with more than 3 dimensions, we can discuss examples in 2 or 3 dimensions and draw analogies to higher dimensional spaces. Figure 4 shows one of many 2-dimensional representations of the norbornane data and it is seen that even this simple plot reveals interesting

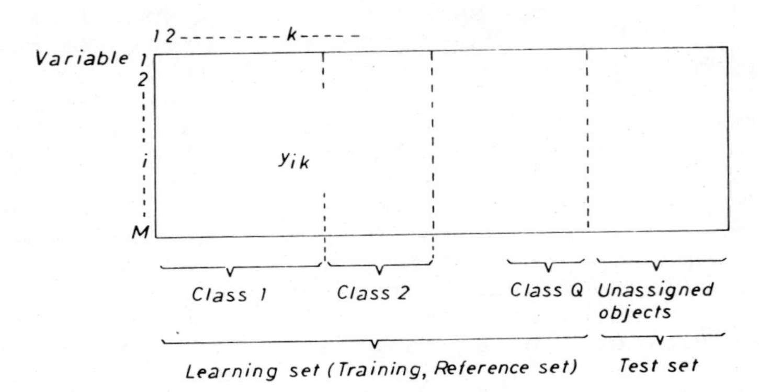

Figure 3. The data matrix in a quantitative analysis of similarity. On each chemical system (object), the values of M variables are measured. The learning set contains objects ordered in classes displaying internal similarity. The test set contains objects of "unknown" class assignment.

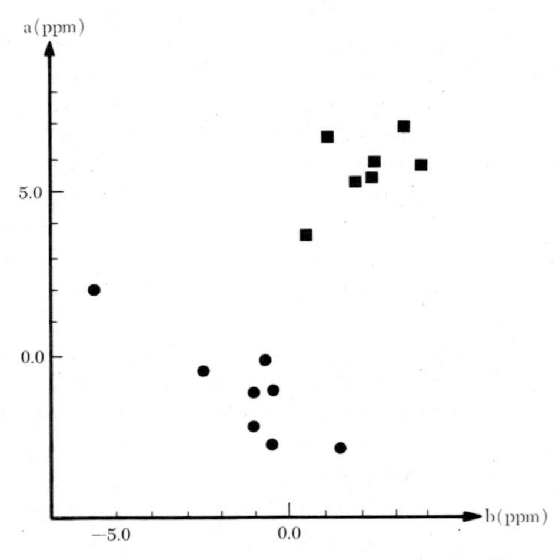

Figure 4. Plot of the difference in relative shift for the C_7 ($\Delta\delta_{7k}$) and C_6($\Delta\delta_{6k}$) carbons against the relative shift for the C_4 carbon ($\Delta\delta_{4k}$) for the compounds 1-15 (see Figure 2). The variables are further rescaled to the same experimental width. Thus $a = (1.43\ \Delta\delta_{7k}\text{-}\Delta\delta_{6k})/\sqrt{2}$ and $b = 2.73\ \Delta\delta_{4k}$. Circles refer to exo *compounds and squares to* endo *compounds. Two well separated classes are obtained.*

regularities within the two classes. Graphical methods will be discussed further in Section 6.8.

5. Which Information Is Wanted from the Data Analysis

For each particular problem there are of course some questions which are specific for the problem. There are usually also a number of questions which are common to all problems of this kind. These questions relate to mainly three types of information;
(*a*) information about each separate class,
(*b*) information about the relations (similarity and dissimilarity) between objects and classes and
(*c*) information about the relations between classes.

5.1 A Single Class.

The information about a single class is based on the analysis of the data observed on the objects "known" to belong to the class (the training, learning, or reference set of the class). The primary goal is usually to get a picture of the "data profile" of the class, i.e., the typical behavior of the objects in the class. In SIMCA, this data profile is described in terms of the parameters $\underline{\alpha}$ and $\underline{\beta}$ in eq. (1) (see section 6.2), which in turn can be seen to define a simple surface (line or hyperplane) in M-space.

Second, it is often of interest which variables that "take much part" in this data profile and which (if any) that are less involved in the class profile. This is further discussed in section 6.5.2.

Third, in the analysis of real data, the learning set of the class is often chosen under some uncertainty. Hence, one is interested to find out whether one or several of the objects in the learning set have deviating properties -- either due to an incorrect class assignment or because of a non-typical behavior -- one wants to find outliers in the training set (see section 6.5.1.).

5.2 Relations Between Objects and Classes.

The class assignment of the objects in the test set in sometimes the primary goal of the data analysis. In chemistry one is usually also interested in other aspects as touched on in sections 5.1 and 5.3. We stress this because some methods used in "pattern recognition" are "optimal" when (*i*) the sole goal is the classification of unassigned objects and (*ii*) one is completely certain that the objects in the training set all are correctly assigned and typical of the classes. When more information is desired (see above) and when the possibility of outliers must be considered, some of this classification optimality must be compromised for other aspects such as information about the data profiles of the classes and outliers.

The information about the relation between the objects and the classes can take different forms. The crudest is that an object either belongs to a class or not. Though sufficient in some cases, it is usually desired to instead get a probability measure of this class assignment. Thus, SIMCA gives, for each object, a probability for each class that the object belongs to the class (see section 6.6). These probabilities need not sum up to one. We can consider a particular object to have very low probabilities to belong to the given "known" classes -- it is an object of "a new kind". We can also have high probabilities for an object to belong to two (or several) classes -- either the data are not sufficient to make an unique assignment, or the object in fact does belong to several classes like a chemical compound having several functional groups or a patient suffering from several diseases.

5.3 Relation Between Classes.

Here we might be interested in whether two of the classes are "close" to each other compared to the other classes -- we desire information about the "distances" between the classes (section 6.5.7). Another often important type of information is the "distance" between two classes for each of the variables; i.e., how important each variable is to distinguish between two classes. When collecting such information for one variable over all class-pairs or, perhaps better, over all "close" class-pairs, we get an idea about the discrimination power of the variable (section 6.5.4). In many problems, the discrimination power of the variables is of great importance for the interpretation of the data structure; in particular when a large number of variables have been included without prior information about their relevance for the problem.

When reducing the number of variables (see section 6.5.6), however, we must not base this reduction on the discrimination power of the variables since this will result in a gross over estimation of the differences between classes. Instead, as discussed below, this selection is better based on the modelling power of the variables which basically is how much each variable "takes part" in the modelling of all the classes instead of just a single class as discussed in section 5.1 (section 6.5.3).

6. The SIMCA Method

The method* is based on modelling each class by a separate model. The main features are described in the following sections. Of those, 6.3 through 6.6 are rather technical in nature and might be skimmed by the reader more interested in the philosophical points.

6.1 Describing a Class by Mathematical Model.

We can see one primary goal of the data analysis as finding "the regular behavior" of the objects in the different classes. The class concept was introduced to simplify the discussion of the behavior of a collection of objects -- instead of discussing each individual object we reduce the complexity by the discussion of a much smaller number of classes. With this philosophical motivation for the classes it is natural to treat each class as independent from the other classes. This corresponds, in mathematical terms, to the construction of one mathematical model for each class.

The first question is whether this is at all possible, after all we know only that the objects in one class are in some way similar. The trick is to use mathematical models which can approximate <u>any</u> regular behavior of a class of similar objects.

Let us, for illustrative purposes for a while look at a simpler case. For bivariate data $y = f(x)$ we know a large number of approximative models. Thus, provided that $f(x)$ has some continuity properties, it can be Taylor expanded around $x = x_o$ giving a polynomical series of arbitrarily good approximating power when sufficiently many terms are included in the expansion.

In the same way, using multidimensional Taylor expansions, it can be shown that under some rather general assumptions, data $\underline{y}_{ik}$ observed of a class of similar objects can be approximated arbitrarily well by the principal components expansion (1) in the next

*
SIMCA is an acronym for Statistical Isolinear Multiple Components Analysis (due to Dave Duewer) <u>or</u> Soft Independent Modelling of Class Analogy <u>or</u> SIMple Classification Algorithm <u>or</u> SIMilarographic Computer Analysis <u>or</u>......

section (5).

The chemist should not be surprised by this result. Simple few term principal components (PC) models have for some time been used to describe chemical data observed on similar objects (usually chemical reactions) under the names of Linear Free Energy Relationships (LFERs) and Extra Thermodynamic Relationships (ETRs) (11,12). The Bronsted and Hammett equations (13,14) are notable examples. The relation between LFERs and chemical pattern recognition has been pointed out by Hammond (15).

The most important condition for these PC models to be able to approximate the data y_{ik} observed on a class of similar objects is that the variables *i* indeed are, at least in part, a realization of this similarity within the class. In other words, the variables chosen to characterize the objects must be relevant to the problem. This condition is, or course, of the same fundamental importance for all methods of data analysis. One advantage with the SIMCA method is that it provides a measure for the relevance of each variable (see sections 6.5.2 and 6.5.4). This gives the opportunity to include variables of doubtful relevance in initial stages of the data analysis, followed by calculation of their actual relevance and the deletion of irrelevant variables.

6.2 Principal Components (PC) Models As Similarity Models.

A data matrix $\underline{Y}^{(q)}$ with the elements y_{ik} can, provided that a few assumptions are fulfilled, be approximated arbitrarily closely by the PC model (5):

$$y_{ik}^{(q)} = \alpha_i^{(q)} + \sum_{a=1}^{A_q} \beta_{ia}^{(q)} \theta_{ak}^{(q)} + \varepsilon_{ik}^{(q)} \quad (1)$$

Here the index *q* indicates that the data belong to class *q*. The assumptions on which this model is based are two; (*i*) the data y_{ik} are generated by a "smooth" process and (*ii*) the cases (objects) with index *k* are "similar".

The first assumption corresponds to an assumption that the function generating the data is several times differentiable so it can be Taylor expanded. This assumption is very natural with chemical data. Whatever detailed theory one has for the processes generating the data y_{ik}, most such theories can be formulated as $\underline{y}$ being solutions to an operator equation which automatically makes this "smoothness" assumption fulfilled.

We note that this first assumption is generally fulfilled when $\underline{y}_{ik}$ are measured data. When $\underline{y}_{ik}$ is of some other kind, say discrete data deduced from a chemical structure such as the presence or absence of a nitro-group in a certain position, this assumption is far from fulfilled. Numerically, the method works anyway, but the user must be cautioned that the interpretation and use of the results is less certain.

The second assumption about similarity is needed to make the number of terms in the Taylor expansions limited, i.e., the number of product terms in model (1), $\underline{A}$, small. This is in a natural way related to how the class defining the choice of the data $\underline{y}_{ik}$ is selected. If it is done in a good way which indeed corresponds to a natural similarity within the class, model (1) will describe data well but if done in a "bad" way, the model (1) will give an ill-defined description.

Geometrically, model (1) corresponds to an A-dimensional hyper plane in M-space. For the simplest cases, A = 0 and A = 1, this corresponds to a single point and a straight line. This makes the graphical illustration of model (1) rather simple (see Figure 7).

Model (1) is related to several mathematical methods finding use in chemistry. LFERs have already been mentioned in the previous section. Ratio-matching (16) is a special case of eq. (1) when the variable means $\underline{\alpha}_i$ are small and the number of product terms $\underline{A}$, is one. The latter is often fulfilled for data observed on closely similar objects. Factor analysis (17,18,19) has a mathematical formulation closely related to eq. (1). The main difference to SIMCA is that one usually in factor analysis assumes that a single model is valid over all data (i.e., a global model), while SIMCA is based on one separate model for each class.

Statistically, the model (1) is closely connected to the models of multiple regression (20). Thus, the values $\underline{\beta}_{ia}$ in eq. (1) correspond to the independent variables X_{ia} in multiple regression. The principal difference is that the latter (X) are known by observation while the former (β) are estimated from the $\underline{N}$ realization vectors $\underline{y}_{ik}$ (k = 1,2,...,N).

Once the values of $\underline{\beta}$ are estimated in SIMCA from the observed values of $\underline{y}_{ik}$ for the training set of the class -- later uses of the model correspond to multiple linear regressions with the values of $\underline{\beta}$ fixed to these estimated values.

Thus, the SIMCA classification of an unassigned object corresponds to multiple regressions, one for each class model (index $\underline{q}$), of the data of the object (denoted by $\underline{y}_{ip}$) on the parameter vectors $\underline{\beta}$

$$y_{ip} - \alpha_i^{(q)} = \sum_{a=1}^{A_q} t_{ap}^{(q)} \beta_{ia}^{(q)} + e_{ip}^{(q)} \qquad (2)$$

The "distance" between the object $\underline{p}$ and the class models is related to the residual standard deviation $s_p^{(q)}$

$$s_p^{(q)} = \left[\sum_{i=1}^{M} \{e_{ip}^{(q)}\}^2 / (M-A_q) \right]^{1/2} \qquad (3)$$

To conclude this section, we state the two main numerical problems involved in the application of model (1) to a given data matrix $\underline{Y}$, obtained by observing the values of M variables on <u>one class</u> of N similar objects (for Q classes, each class is treated in the same way, giving Q independent class models).

(i) Estimation of the number of product terms in eq. (1), $\underline{A}$. The estimation of this number corresponds to the separation of the data Y into signal -- the parameters $\underline{\alpha}$ and $\underline{\beta}$ -- and noise ($\underline{\varepsilon}_{ik}$). We realize that a large $\underline{A}$ gives small residuals ($\underline{\varepsilon}$) -- much signal and little noise, and the reverse. Hence it is of utmost importance to get a good value of A, a value too large will give too much apparent structure in the data leading to too far-reaching conclusions. A value of A too small will correspond to an under-utilization of the information actually contained in the data.

(ii) Once the values of A are estimated for the class, the estimation of the parameters $\underline{\alpha}$ and $\underline{\beta}$ in the model (1) is numerically rather trivial. This is further discussed in section 6.3.

6.3 Parameter Estimation in the PC Model (1).

For a given A, the estimation of the parameters $\underline{\alpha}$ and $\underline{\beta}$ in eq. (1) can be accomplished by several numerical methods. This problem is in mathematics called by many names; singular value decomposition, eigenvector decomposition, Karhunen-Loève expansion, matrix diagonalization and other names. In SIMCA we have, for different reasons, used a special type of algorithm called a NIPALS algorithm (21,22). This works iteratively, determining one vector β_{a+1} after another (β_a) using the fact that these vectors are mutually orthogonal. This iterative "peeling" procedure gives two advantages; it is efficient in the cross-validatory estimation of A (see next section) and it can be adapted to handle incomplete data matrices (some observations are missing).

6.4 The Number of Components (A) in PC Models.

As discussed in section 6.2, it is essential not to overestimate or underestimate the amount of information contained in the data matrix Y. This amount is directly related to the number of terms in model (1), i.e., A. We have used the technique of double cross-validation which gives the value of A corresponding to the optimal predictability of the data set back on itself using model (1). This takes some explanation.

Assume that we have estimated the parameters α_i and β_{ia}, θ_{ak} for a = 1,2 up to A for a given data matrix Y of a particular class. We can then see if the addition of another term $\beta_{i,(A+1)}\ \theta_{(A+1),k}$ betters the fit of model (1) to the data Y in the following way.

1. Calculate the residuals after A terms

$$\varepsilon_{ik}^{(q)} = y_{ik}^{(q)} - \alpha_i^{(q)} - \sum_{a=1}^{A_q} \beta_{ia}^{(q)}\ \theta_{ak}^{(q)} \qquad (4)$$

2. Delete part of the matrix elements in $\{\varepsilon_{ik}^{(q)}\}$ and fit the model to the remaining elements

$$\varepsilon_{ik}^{(q)} = \beta_{i,(A+1)}^{(q)}\ \theta_{(A+1),k}^{(q)} + \delta_{ik}^{(q)} \qquad (5)$$

Since the NIPALS method (21,22) works also with incomplete data, this fitting is easily accomplished

3. For the deleted elements (denoted by ε_{ik}^{*}), calculate the prediction error Δ_{ik}

$$\Delta_{ik}^{(q)} = \varepsilon_{ik}^{*\ (q)} - \beta_{i,(A+1)}^{(q)}\ \theta_{(A+1),k}^{(q)} \qquad (6)$$

4. Restore the matrix $\{\underline{\varepsilon}_{ik}^{(q)}\}$ and then delete other elements and repeat steps 4 and 5 until all elements of $\{\underline{\varepsilon}_{ik}^{(q)}\}$ have been deleted once.

5. Form the sum of all prediction errors in square

$$D = \sum_{i}^{M} \sum_{k}^{N_q} \Delta_{ik}^{(q)2} \qquad (7)$$

Compare D with the sum of residual squares $S^{(q)}$

$$S^{(q)} = \sum_{i}^{M} \sum_{k}^{N_q} \varepsilon_{ik}^{(q)2} \qquad (8)$$

If the ratio between $\underline{D}$ and $\underline{S}$ exceeds one, the $(A+1)^{st}$ product term has not bettered the prediction error of model (1) and we conclude that $\underline{A}$ terms were sufficient.

If D/S is less than one, we make a components analysis of the full matrix $\{\underline{\varepsilon}_{ik}\}$, subtract the $(A+1)^{st}$ term and start again at step $\overline{1}$, but with A = A+1.

The details of this cross-validation methodology will be published elsewhere (23); it suffices here to say that the method works very well with both simulated and real data.

If the values of $\underline{A}_q$ for the different classes come out very different (differ by more than two) one should use different values of $\underline{A}_q$ for the different classes. If they agree within 1, however, it is practical to use the same value of $\underline{A}_q$ for all the classes.

6.5 Information Contained in the Residuals.

The SIMCA method is based on the least squares framework. Thus the parameters $\underline{\alpha}$, $\underline{\beta}$ and $\underline{\theta}$ in eq. (1) are calculated to minimize the sum of squared residuals ($\underline{\varepsilon}_{ik}$). This estimation is consistent for most probable distributions of $\underline{\varepsilon}_{ik}$ but in order for the methods below to be efficient, the residuals should be approximately normally distributed. Hence, this should be tested for. This is easiest done by making histograms of $\underline{\varepsilon}_{ik}$ for each variable and class. If the residuals are highly non-normal, the data should be transformed, the reader is referred to Box and Cox (24) for a discussion of the problem.

When the number of objects in each class is small, however, (smaller than 20-30) any test of normality is highly uncertain and crude and in this typical case one can only hope that normality is approximately fulfilled.

The size of the residuals after fitting data of a class reference set or data of individual objects to the various class models reveals interesting pieces of information. This will be discussed after the definition of suitable measures of "residual size".

Besides equations (1) through (3) we shall need, in the treatment below, the following quantities:

The total residual standard deviation of class q; $s_o^{(q)}$ (ε_{ik} from eq. 1):

$$s_o^{(q)} = \left[\sum_i \sum_k \varepsilon_{ik}^{(q)2} / (N_q - A_q - 1)\,(M-A_q) \right]^{1/2} \qquad (9)$$

The residual standard deviation of variable i over all the data in the training set

$$s_i = \left[\frac{1}{Q} \cdot \sum_{q=1}^{Q} \frac{M}{(M-A_q)} \sum_{k=1}^{Q} \varepsilon_{ik}^{(q)2} / N_q-A_q-1 \right]^{1/2} \qquad (10)$$

The corresponding standard deviation of the training set data y:

$$s_{i,y} = \left[\sum_{q=1}^{Q} \sum_{k=1}^{N_q} (y_{ik}^{(q)} - \bar{y}_i)^2 / (\sum_{q=1}^{Q} N_q) -1 \right]^{1/2} \qquad (11)$$

The average of y_{ik} over the training set

$$\bar{y}_i = \sum_{q=1}^{Q} \sum_{k=1}^{N_q} y_{ik}^{(q)} / \sum_{q=1}^{Q} N_q \qquad (12)$$

The residual standard deviation of the objects in class r when fitted to the class model q ($e_{ip}^{(q)}$ from eq. 2: y_{ip} then denotes data belonging to class vectors r)

$$s_r^{(q)} = \left[\sum_i^{M} \sum_{p=1}^{N_r} e_{ip}^{(q)2} / (M-A_q) \cdot N_r \right]^{1/2} \qquad (13)$$

The corresponding standard deviation of variable $\underline{i}$

$$s_{i,r}^{(q)} = \left[\frac{M}{(M-A_q)} \sum_{p=1}^{N_r} e_{ip}^{(q)2} / N_r \right]^{1/2} \quad (14)$$

6.5.1 Outliers in a Class Reference Set.

The size of the residuals for a particular object (denoted by $\underline{\varepsilon}_{ip}$) can be compared with the "normal" size of the class by means of, say, an F-test. If this comparison shows that the residuals of the object are large, this indicates that the object is an outlier, either because it was incorrectly assigned to the class or because it shows an atypical behavior. Thus, the test is based on the ratio

$$F = s_p^{(q)2} \cdot \frac{N_q}{(N_q-A_q-1)} / s_o^{(q)2} \quad (15)$$

The standard deviations are defined in eq. (3) and (9). The correction $N_q/(N_q-A_q-1)$ is needed because the vector $\underline{p}$ has been included in the computation of the class parameters $\underline{\alpha}$ and $\underline{\beta}$.

The F-value in eq. (15) is compared with critical F-values with $(M-A_q)$ and (N_q-A_q-1) $(M-A_q)$ degrees of freedom.

If outliers are found in the training set, they should, as long as they are not too many (say not more than 10% of the training set) be deleted and the class parameters should be recalculated.

6.5.2 The Modelling Power of a Variable.

In the method like SIMCA where each class is described by a mathematical model, measures of how much of the variation in the original data that are described by the models are of great interest. These measures are simplest obtained by comparing residual standard deviations with the corresponding data standard deviations. Thus we can calculate the modelling power of a variable $\underline{i}$ over all classes as:

$$\psi_i = 1 - (s_i / s_{i,y}) \quad (16)$$

Here s_i and $s_{i,y}$ are defined in eq. (10) and (11) respectively.

A value of ψ_i close to one indicates a high and close to zero a low modelling power.

6.5.3 The Data Relevance for a Class.

Similarly to the modelling power for a single variable (previous section) we can calculate how much of the variation in the data of a class q is described by the model:

$$\psi_o^{(q)} = 1 - s_o^{(q)} / s_y^{(q)} \qquad (17)$$

Here, $s_o^{(q)}$ is given in eq. (9) and $s_y^{(q)}$ is the standard deviation of the data in the class (y_i from eq. 12):

$$s_y^{(q)} = \left[\sum_{i=1}^{M} \sum_{k=1}^{N_q} (y_{ik}^{(q)} - y_i^2) / (M \cdot N_q) \right]^{1/2} \qquad (18)$$

The average of $\psi_o^{(q)}$ over all classes gives a measure of the total data relevance.

6.5.4 Discriminatory Power of a Variable.

By comparing the fit of objects in two given classes q and r to (a) their own class models and (b) to the other class model (objects in class q to model r and vice versa) we can get an idea of the distance between the two classes over all variables (section 6.5.8) and also for each variable. Using the residual standard deviations in eq. (14) we can calculate the discriminatory power of variable i between the two classes q and r as:

$$\Phi_i^{(r,q)} = \left[\frac{s_{i,r}^{(q)2} + s_{i,q}^{(r)2}}{s_{i,r}^{(r)2} + s_{i,q}^{(q)2}} \right]^{1/2} - 1 \qquad (19)$$

A value of $\Phi_i^{(r,q)}$ close to zero indicates a low discriminatory power and much above one a good power.

When more than two classes are present, an aggregate measure is obtained by summation over all class-pairs or just over all close class pairs, n_{pair} according to the distances discussed in section 6.5.7).

$$\Phi_i = \sum_{r \neq q} \sum \Phi_i^{(r,q)} / n_{pair} \tag{20}$$

6.5.5 Weighting of Variables.

It is customary to scale the variables so that they all have the same variance over the training set (auto-scaling) (9). This is to avoid an exaggerated influence of a variable showing large variation in the original data.

After the first stage of the analysis, the measures of modelling power and discrimination power give additional information about the relevance of the variables for the problem. Hence, in later stages, the variables might, if so warranted, be weighted by a multiplication by their modelling power.

Such a weighting usually has only a small effect on the analysis, however, and it is usually better to make a zero-one weighting as discussed in next section.

6.5.6 Selection of Variables from Many.

When a large number of variables have been included, a number of these often turn out to have a low modelling power (ψ_i smaller than, say, 0.3 in eq. 16). Such variables are then best completely deleted and the computations redone for the reduced data set. We should note that variables must not be deleted solely on the basis of their discriminatory power since this leads to a gross exaggeration of the differences between classes; the design of a method looking for differences must not be conditioned to maximize such differences. A reasonable compromise can be reached, however, by deleting only such variables which have both low modelling and low discrimination power.

6.5.7 Distance Between Two Classes.

In the same way as the discrimination power for single variables were calculated in section 6.5.4, we can calculate a distance between two classes using the residual standard deviations over all variables.

$$D^{(r,q)} = \left[\frac{s_r^{(q)2} + s_q^{(r)2}}{s_o^{(q)2} + s_o^{(r)2}} \right]^{1/2} - 1 \tag{21}$$

A distance $D^{(r,q)}$ close to zero indicates that the two classes r and q are virtually identical (with respect to the given data) and values larger than one indicate real differences. A quantitative test on the significance of $D^{(r,q)}$ is obtained by using the fact that $(D^{(r,q)} + 1)^2$ is approximately F - distributed, with $(N_r - A_q)(M-A_q) + (N_q-A_r)(M-A_r)$ and $(N_r-A_r-1)(M-A_r) + (N_q-A_q-1)(M-A_q)$ degrees of freedom.

6.6 Classification of New Objects (Test Set).

In order to assign objects in the test set to the classes, each of these objects is fitted to each class model according to eq. (2). The residual standard deviation (eq. 3) corresponds to the orthogonal distance between object p and class q (Figure 5). However, when calculating the parameters $\theta_{ak}^{(q)}$ of the q.th class one also obtains information about the normal range of $\theta_{ak}^{(q)}$ for the class.

If an object p, when fitted to the class model (eq. 2), gets coefficients t_a falling outside the normal range of θ_a for the class, one instead calculates the "distance" between object p and class q as $d_p^{(q)}$ (Figure 5)

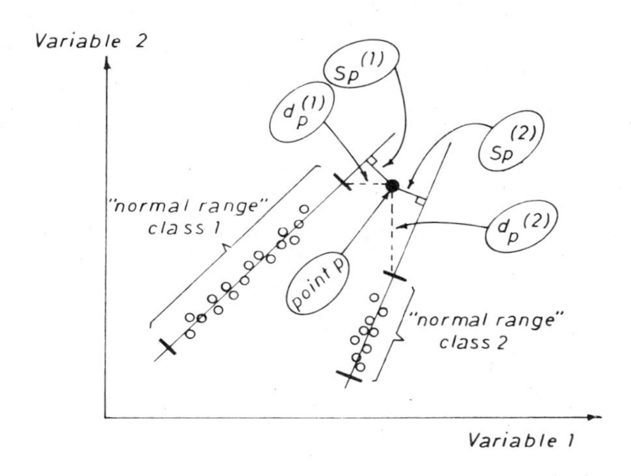

Figure 5. When a point p *gets coefficients* t_{ap} *(eqn. 2) outside the "normal range" of a class* q *(eqn. 23, 24), the distance between the object* p *and the class is measured by* $d_p^{(q)}$ *(eqn. 22).*

$$d_p^{(q)} = \left[s_p^{(q)2} + \Sigma\, \Phi_a^2 (t_a - \theta_{a,lim}^{(q)})^2 \right]^{1/2} \qquad (22)$$

Here the summation is made only over terms where $\underline{t}_a$ is outside the normal range $\underline{\theta}_{a,low}^{(q)}$; $\underline{\theta}_{a,high}^{(q)}$

This range is defined by the minimal and maximal values of $\underline{\theta}$ plus minus $\underline{t}/2$ standard deviations of $\underline{\theta}$ where $\underline{t}$* is the $\underline{t}$ distribution with N_q degrees of freedom

$$\theta_{a,lim}^{(q)} = \begin{cases} \theta_{a,max}^{(q)} + t^*/2 \cdot s_{\theta,a}^{(q)} & (23) \\ \theta_{a,min}^{(q)} - t^*/2 \cdot s_{\theta,a}^{(q)} & (24) \end{cases}$$

$$s_{\theta,a}^{(q)} = \left[\sum_{k=1}^{N_q} \theta_{ak}^{(q)2} / N_q \right]^{1/2} \qquad (25)$$

The coefficient $\underline{\Phi}_a$ in eq. (22) is introduced to make $\underline{s}_p^{(q)2}$ and $(\underline{t}_a - \underline{\theta}_{a,lim})^2$ comparable

$$\Phi_a = s_p^{(q)} / s_{\theta,a}^{(q)} \qquad (26)$$

The object $\underline{p}$ is then assigned to the class showing the best fit, i.e., the smallest $\underline{d}_p^{(q)}$, provided that this $\underline{d}_p^{(q)}$ is not much larger than the standard deviation of the class according to an F-test $(M-A_q)$ and (N_q-A_q-1) $(M-A_q)$ degrees of freedom:

$$F = d_p^{(q)2} / s_o^{(q)2} \qquad (27)$$

Furthermore, a unique assignment is obtained only when the ratio between the next smallest (here denoted by $S_p^{(r)}$) and the smallest residual variances ($S_p^{(q)}$) is larger than a critical F-value with $M-A_r$ and $M-A_q$ degrees of freedom

$$F = d_p^{(r)2} / d_p^{(q)2} \qquad (28)$$

To conclude: the classification of an object p in the test set can give several different results.

a. Object p is uniquely assigned to class q. It fits the closest class q within the typical standard-deviation of that class (F in eq. (27) is smaller than the critical value). In addition, the distance to the next closest class is much larger according to the F-test in eq. (28).

b. Object p fits several classes $q_1, q_2 \dots q_n$. It fits each of these classes within the typical standard deviations of the classes (eq. 9), but the ratios between the distances to the classes $q_1 \dots q_n$ are not different according to eq. (28). The reason for this ambiguity can be two; either the data are not sufficient to distinguish between the different classes with respect to object p (more variables with better discrimination power then be measured) or the object actually belongs to several classes being either a border-line case or having properties of several classes like a compound with several functional groups.

c. Object p fits no class within its typical standard deviation (eq. 9). We then conclude that the object is of a new, hitherto not see, type, a member of a new class.

6.7 Missing Data.

The presentation has so far been implicitly based on the assumption that the data matrix of each class reference set has been complete, i.e., all elements $y_{ik}^{(q)}$ are defined by measured values. In many chemical problems, however, the class data matrices are incomplete due either the difficulty to measure all values, or the absence of a defined value for some variables and objects, e.g., a compound may have an ill-defined melting point or a certain peak might be missing in the IR spectrum of a compound.

The model (1) may still be fitted to an incomplete data matrix using an extended NIPALS method (23). All degrees of freedom in the expressions for the various residual standard deviations must then be changed to take missing observations into account, but that presents no problem, 1 is subtracted from the denominators for each observation missing in the calculation.

The determination of the number of product terms in the models (section 6.4) can hopefully also be modified to cope with missing data. Presently, however, with missing data one must use a standard value of $\underline{A}_q$. We recommend using $\underline{A}_q$ = 2 or 3 unless M is smaller than 10 when $\underline{A}_q$ should be smaller than M/4 ($\underline{A}_q$=1 is always admissible, however, even if M=2 or 3).

6.8 Graphical Methods.

The model (1) has a simple representation in M-space being a line ($\underline{A}$=1) or an $\underline{A}$-dimensional hyperplane. In cases when $\underline{A}$=1 gives a good representation of the data, this can be seen also in linear projections of M-space down on two-space. Such projections will project also the line of the class model to a line in the 2-dimensional space. This is the basis for the following graphical method which can easily be applied with the help of graph paper and a calculator when the number of variables M is smaller than 5 or 6 (below we illustrate the method with M=4).

a. Scale the variables by dividing them by their range over all data.

b. Divide the variables "randomly" into two groups, say group A containing variables 1 and 4 and group B containing variables 2 and 3.

c. Plot, for each object, the sum of the variables in group A against the sum of the variables in group B, i.e.

$(y_{1k}^* + y_{4k}^*)$ against $(y_{2k}^* + y_{3k}^*)$ the asterisks denoting that the y-values are scaled.

d. Plot instead the differences between the variables in groups A against the differences between the variables in group B, i.e., $(y_{1k}^* - y_{4k}^*)$ against $(y_{2k}^* - y_{3k}^*)$.

e. Redistribute the variables creating two new groups A' and B', say A'=variables 1 and 3 and B'=variables 2 and 4.

f. Go through steps c and d for these two new groups.

In this way one obtains four different plots of the data and patterns that are consistent in two or three of the plots are indications of real regularities also in M-space.

As with all methods based on projections, it is important to realize that a single plot might reveal apparent regularities which in fact are a result of the choice of projections. One must therefore always use at least two different projections as graphical illustrations of data and results.

When a computer is used, a large number of different projections can be generated more or less automatically. The reader is referred to Kowalski (25) for an illuminating discussion.

6.9 The Norbornane Example.

The analysis of the norbornane data gave the following results, the chemical significance of which is discussed in next section.

0. The data were first autoscaled by subtracting the variable means and dividing by 4.69 times the standard deviations (bottom of Table I). All the results refer to the analysis of the scaled data.

1. The dimensionality of the data was found by applying crossvalidation separately to the two class matrices defined by the data in Table I. This shows that each class is best described with model (1) having $A=1$.

2. Fitting model (1) to the two class matrices gives the parameter values of the two classes shown in Table II for the three variables containing any information (see below). Variables with low discrimination and modeling power were step wise deleted from further analysis and the computations were redone for the reduced data matrices (see Table III).

3. The data in the learning set were fitted to the two class models (3 variables) giving the results in Table IV.

4. The data in the test set were fitted to the two class models (3 variables) giving the results in Table V.

6.9.1 Classification of Exo and Endo 2-Substituted Norbornanes.

The results from the SIMCA classification show that only the so-called γ-carbons (C_4, C_6 and C_7) contain information whether a compound is an endo or exo 2-substituted norbornane. Thus the shift differences observed for the α, β and Δ carbons contain only

Table I. Relative ^{13}C NMR shifts $\Delta\delta$ in ppm for the C_1-C_7 carbons in the norbornane framework[a]. The $\Delta\delta$ values are obtained as shift differences between the actual structure and the unsubstituted norbornane (X=H)

Comp. no.	substituent	class q	$\Delta\delta_{1k}$	$\Delta\delta_{2k}$	$\Delta\delta_{3k}$	$\Delta\delta_{4k}$	$\Delta\delta_{5k}$	$\Delta\delta_{6k}$	$\Delta\delta_{7k}$
1	$-CH_3$(exo)	1	6.7	6.7	10.1	0.5	0.2	-1.1	-3.7
2	$-NH_2$(exo)	1	8.9	25.3	12.4	-0.4	-1.2	-3.1	-4.4
3	-OH(exo)	1	7.7	44.3	12.3	-1.0	-1.3	-5.2	-4.1
4	-F(exo)	1	5.6	65.8	10.0	-1.9	-1.8	-7.5	-3.4
5	-CN(exo)	1	5.5	1.0	6.3	-0.3	-1.5	-1.6	-1.3
6	-COOH(exo)	1	4.6	16.7	4.4	-0.2	-0.3	-1.0	-1.8
7	$-CO_2CH_3$(exo)	1	5.1	16.4	4.2	-0.4	-1.1	-1.4	-2.1
8	$-CH_2OH$(exo)	1	1.8	15.1	4.4	-0.2	0.2	-0.7	-3.3
9	$-CH_3$(endo)	2	5.4	4.5	10.6	1.4	0.5	-7.7	0.2
10	$-NH_2$(endo)	2	6.8	23.3	10.5	1.2	0.6	-9.5	0.3
11	-OH(endo)	2	6.3	42.4	9.5	0.9	0.2	-9.7	-0.9
12	-CN(endo)	2	3.4	0.1	5.5	0.2	-0.7	-4.9	0.0
13	-COOH(endo)	2	4.2	16.2	2.1	0.9	-0.6	-4.8	1.9
14	$-CO_2CH_3$(endo)	2	4.0	15.9	2.2	0.7	-0.7	-5.0	1.7
15	$-CH_2OH$(endo)	2	1.7	12.8	4.0	0.4	0.2	-7.2	1.4
mean value $\Delta\delta_{ik}$			5.37	15.8	8.02	0.25	-0.36	-3.65	-1.22
standard dev. σ_i			1.68	15.8	3.15	0.78	0.75	3.44	2.31

[a] Data from Grutzner, J. B., Jautelat, M.; Dence, J. B., Smith, R. A., and Robert J. D.; J. Amer. Chem. Soc. (1970), 92, 7107.

Table II. Parameters for the exo and endo class models

		$\alpha_i^{(q)}$	$\beta_{i1}^{(q)}$
class 1 (q=1)	i=4	-0.201	0.775
	i=6	0.058	0.613
	i=7	-0.165	0.149
class 2 (q=2)	i=4	0.153	0.495
	i=6	-0.206	-0.764
	i=7	0.172	-0.415

Table III. Discrimination (Φ) and modelling (Ψ) power (see section 6.5.3 and 6.5.5) for the norbornane carbons as a function of the number of variables fitted to eq. (9). Values are shown for seven and three variables.

	i=1	i=2	i=3	i=4	i=5	i=6	i=7
Φ_i	0.4	1.	1.	14.	5.	46.	20.
Ψ_i	0.23	0.25	0.45	0.36	0.50	0.60	0.55
Φ_i				22.		69.	23.
Ψ_i				0.72		0.74	0.50

Table IV. The result of the SIMCA classification of the compounds 1-15 in the learning set and with A=1 in eq. (1) Residual variance class 1 $\sigma_1^2 = 0.012$ and class 2 $\sigma_2^2 = 0.012$. F-test according to eq. (15).

Compound no.	known class	standard deviation (closest class)	standard deviation (next closest class)	F[a] (closest class)
1	1	0.10 (1)	0.37 (2)	1.1 (1)
2	1	0.10 (1)	0.24 (2)	1.1 (1)
3	1	0.06 (1)	0.46 (2)	0.4 (1)
4	1	0.03 (1)	0.57 (2)	0.1 (1)
5	1	0.11 (1)	0.25 (2)	1.3 (1)
6	1	0.07 (1)	0.28 (2)	0.5 (1)
7	1	0.07 (1)	0.31 (2)	0.5 (1)
8	1	0.05 (1)	0.37 (2)	0.3 (1)
9	2	0.08 (2)	0.44 (1)	0.7 (2)
10	2	0.03 (2)	0.48 (1)	0.1 (2)
11	2	0.07 (2)	0.43 (1)	0.6 (2)
12	2	0.12 (2)	0.25 (1)	1.4 (2)
13	2	0.09 (2)	0.38 (1)	0.9 (2)
14	2	0.04 (2)	0.36 (1)	0.2 (2)
15	2	0.09 (2)	0.39 (1)	0.9 (2)

[a] $F_{crit}=4.0$

Table V. The result of the SIMCA classification of the compounds 16-43 in the test set. F-test according to eq. (27).

Compound no.	known class	standard deviation (closest class)	F[a] (closest class)	standard deviation (next closest class	F[a] (right class if not closest)
16	1	0.13 (1)	1.3 (1)	0.42 (2)	
17	2	0.13 (2)	1.3 (2)	0.27 (1)	
18	1	0.09 (1)	0.6 (1)	0.40 (2)	
19	2	0.11 (2)	1.0 (2)	0.43 (1)	
20	1	0.08 (1)	0.5 (1)	0.41 (2)	
21	2	0.07 (1)	0.5 (1)	0.26 (2)	5.6 (2)
21'	2	0.10 (2)	0.8 (2)	0.25 (1)	
22	1	0.08 (1)	0.5 (1)	0.36 (2)	
23	2	0.20 (1)	3.2 (1)	0.33 (2)	9.1 (2)
23'	2	0.10 (2)	0.8 (2)	0.33 (1)	
24	1	0.08 (1)	0.5 (1)	0.41 (2)	
25	2	0.09 (2)	0.6 (2)	0.45 (1)	
26	1	0.06 (1)	0.3 (1)	0.38 (2)	
27	2	0.09 (2)	0.6 (2)	0.30 (1)	
28	1	0.09 (1)	0.5 (1)	0.42 (2)	
29	2	0.19 (2)	3.0 (2)	0.35 (1)	
30	1	0.10 (1)	0.8 (1)	0.45 (2)	
31	2	0.30 (2)	7.7 (2)	0.39 (1)	
32	1	0.09 (1)	0.7 (1)	0.40 (2)	
33	2	0.04 (2)	0.2 (2)	0.29 (1)	

Table V. (continued)

34	1	0.14 (1)	1.5 (1)	0.44 (2)	
35	2	0.30 (2)	7.8 (2)	0.37 (1)	
35'	2	0.15 (2)	2.0 (2)	0.48 (1)	
36	1	0.11 (1)	0.9 (1)	0.41 (2)	
37	2	0.07 (2)	0.5 (2)	0.45 (1)	
38	1	0.09 (1)	0.6 (1)	0.41 (2)	
39	2	0.05 (2)	0.2 (2)	0.36 (1)	
40	1	0.18 (2)	2.7 (2)	0.29 (1)	7.0 (1)
41	2	0.32 (2)	8.7 (2)	0.50 (1)	
42	1	0.10 (2)	0.8 (2)	0.32 (1)	9.5 (1)
43	2	0.40 (1)	13.1 (1)	0.44 (2)	16.1 (2)

[a] $F_{crit}=4.0$

individual effects of the substituent but no specific information whether the substituent is situated in *exo* or *endo* postion. The finding that models with dimensionality $A=1$ describes the data well shows that the shift parameters consist of an individual substitute effect $\theta_k^{(q)}$ in addition to the *exo*/*endo* effects $\alpha_i^{(q)}$.

The classification can be visualized by the earlier described plotting method (see section 6.8). In this case two well separated classes are obtained by plotting the difference between the relative shift for C_7 and C_6 against the relative shift for C_4. The variables were rescaled to 2.73 C_4, 1.0 C_6 and 1.43 C_7, where the coefficients are obtained by dividing the experimental width for each variable by the variable with minimum width observed (C_6). (See Figures 4,7 and 8.)

The obtained parameters from the test sets can be used to clarify if other _exo_ or _endo_ 2-substituted compounds closely related to the norbornanes behave in a similar way. Therefore a number of such compounds were chosen to form a test set (compounds 16-43, see Figure 6). The autoscaled shift differences were fitted to the parameters obtained for the training sets (α and β are fixed to the values obtained in the analysis of the reference sets). The goodness of fit of the nonclassified objects to the class models can be estimated by F-test by comparing the residual variances for these objects with the residual variances of the reference sets. (See Table V.) Most of the structures are well-classified i.e., they have the same variance as observed for the objects in the reference sets, a fact which strengthens the proposed assignments in most cases. (See Figure 7.) However, some exceptions can be noted (Figure 8).

(a) The tentative assignment given for the C_4 and C_2 carbons in structure 23 does not provide a correct classification. The reversed assignment for C_4 and C_2 classified this structure well.

(b) For compound 21 two different assignments have been proposed for the C_5 and C_7. One of these classified 21 correctly.

(c) Applying the proposed assignment for 35 this compound fit neither class. However, after reversing the assignment for C_2 and C_4 the compound fit nicely to the _endo_ class.

(d) A group of compounds the terpenes 40-43 have been found not to be well-classified with the parameters from the test set, showing that compounds with methyl groups on the C_7 don't behave in the same way as the shifts for the compounds in the test-set.

7. Applications

The SIMCA-methodology is rather new and the number of published applications therefore is small. Most of the early applications concern structure-reactivity relations of the Hammett-equation type with the data analysis of a single class (_12_,_26-28_). A single class

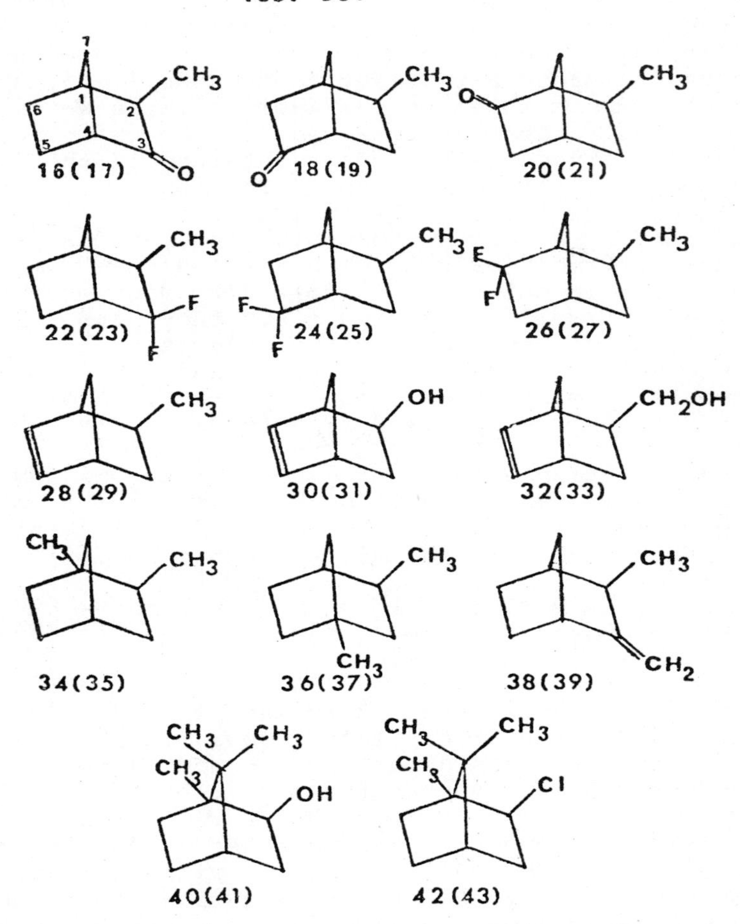

Figure 6. The test set of exo *and* endo *norbornanes and related compounds. Even numbers refer to the* exo *compounds, the odd numbers to the* endo *compounds.*

application in Analytical Chemistry addresses the comparison of analytical methods with respect to their performance on real data (29).

The earliest use of SIMCA in pattern recognition with several classes was made by Duewer, Kowalski and Schatzki (30) who analyzed the 22-dimensional data of 40 classes of simulated oil-spills. They showed that it is possible to determine the source of an oil-spill on the basis of such data. This was followed by the test of SIMCA on the classical Iris data of Fisher (5) and the analysis of spectral data (IR and UV) of two classes of unsaturated carbonyl compounds (12).

Sjöström and Edlund analyzed the ^{13}C NMR data of norbornanes and chlorinated biphenyls (4). The norbornane analysis has been used in the present article as an illustration, Strouf and Wold have used SIMCA to try to find regularities within the two classes of stable and unstable complex hydrides (31). Using structural features such as the size and electronegativity of ligands and central atoms, a 75% prediction rate of the stability of the complexes was achieved, a rate which is significantly better than chance.

The similarity model (1) has also been used in cluster analysis in the grouping of liquid phases for gas chromatography on the basis of retention data of 10 standard compounds (32).

We will not here in detail discuss areas where SIMCA has possible applications. Chemistry is full of problems which in a natural way can be formulated in terms of similarity assessment and/or classification. Suffice it here to mention structural determination of unknown compounds on the basis of spectra, source determination of samples (oil-spills, blood-stains, paper-samples), type assessment of biological individuals (chemotaxonomy) relation between structure and reactivity and -- exciting but difficult -- the efforts to find relations between chemical structure and biological activity (mutagenicity, toxicity, etc.).

8. Future Developments

The main lines of development can be seen for the SIMCA methodology. The first is to extend the similarity models beyond the simple PC models of eq. (1) to cope with particular types of problems where the PC model is less efficient. One example is when one wishes to introduce causality in the mode. Take, for instance, the norbornane application discussed above, and let us post the problem how the variation of ^{13}C NMR spectra within each class depends on the structure

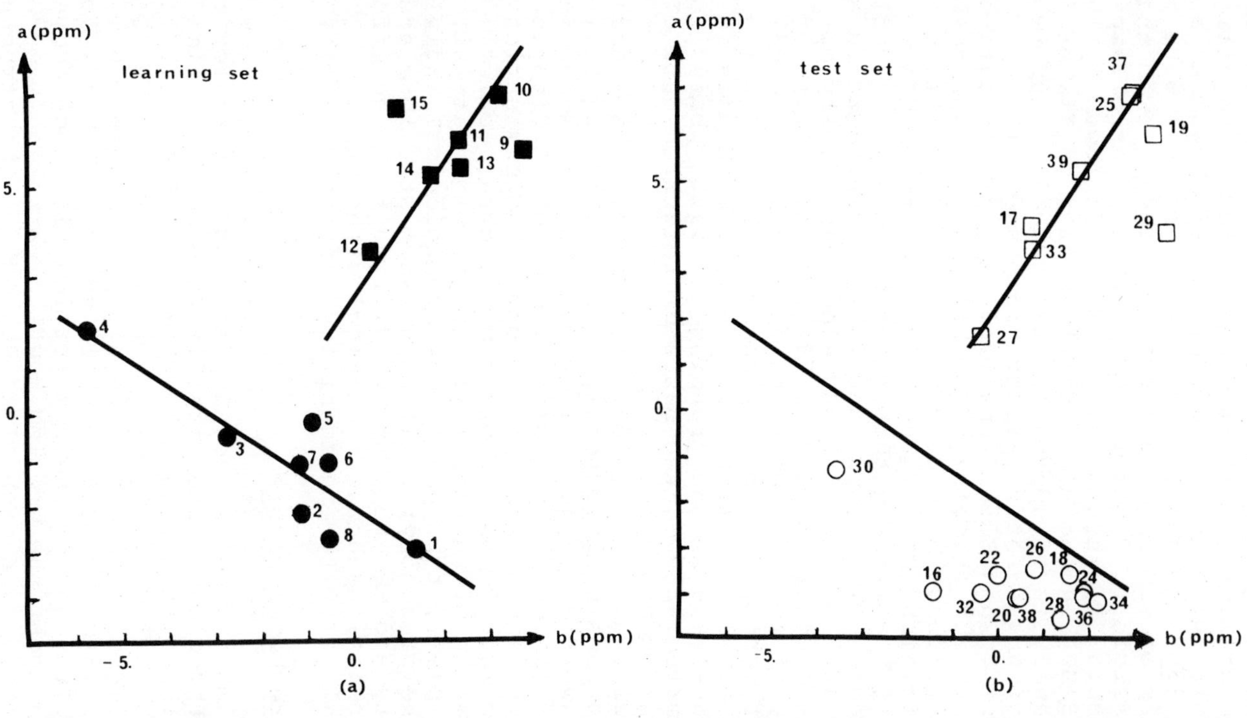

Figure 7. Classification of exo *and* endo *2-substituted norbornanes and similar compounds* $a = (1.43\ \Delta\delta_{7k}\text{-}\Delta\delta_{6k})/\sqrt{2}$ *and* $b = 2.73\ \Delta\delta_{4k}$ (see also *Figure 4*). *(a) The two reference sets (learning set) which are approximately described by straight lines. (b) A test set which is well classified by the learning set.*

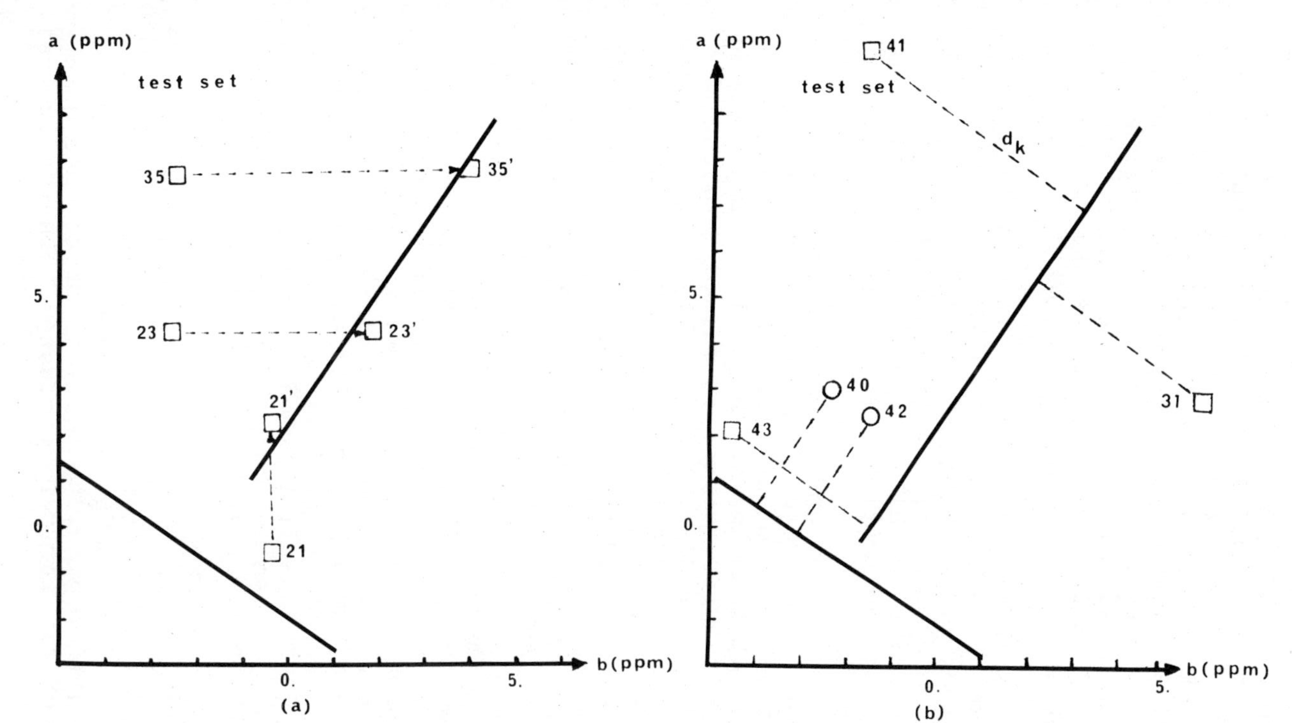

Figure 8. (a) Classification of the endo *compounds 21, 23, and 35 with two different assignments of their* [13]*C NMR spectra. The reversed spectra are denoted with 21′, 23′, and 35′. Compare results in Table V. (b) The terpenes 40–43 and compound 31 are not well-classified by the learning set.* d_k *(broken line) is the difference between an observation and the lines equation for the corresponding class. The normalized distance* $d_k/\sqrt{(M-1)}$ *(M number of variables) is significantly larger than the typical normalized distance (residual standard deviation for the class) for the observations in the reference set. Compare F-test Table V.*

of the compounds. If we describe each structure by means of a number of structural descriptors, say, the size, electronegativity and Taft σ-value of the 12 substituents in positions 1 through 7 (including hydrogen), we have 36 structural descriptors for each compound (block 1). These, together with the 7 NMR-shifts (block 2) give 43 variables for each compound. However, we now wish to analyze the data in terms of cause and effect, we wish to "explain" the variation of the latter 7 variables (block 2) as the "effect" of the variation of the former 36 (block 1). This can be done by expanding the model (1) to two PC-models -- one for each block of variables -- plus a multiple relation between the θ-vectors in block 1 and the θ-vectors in block 2. Such "path" models which currently are being developed for use in econometrics, sociology and other social sciences (33,34) will have many interesting applications also in the analysis of chemical data.

The second line of development involves the application of the SIMCA methodology to data where the grouping (classes) is not initially known. Data-analysis aimed at finding the "natural" grouping of a number of objects with respect to multidimensional data measured on the objects is usually called cluster analysis. A SIMCA cluster analysis would thus try to group the objects so that within each of the resulting groups (clusters) the data would be well described by model (1) (or one of its path-model extensions). Mathematically and computationally, this is a much more difficult problem than the one where the classes are defined à priori but the results are also more rewarding in terms of new knowledge.

Possible chemical applications of cluster-analysis are as many as those of classification. Any large collection of chemical systems, say solvents, reactions of a certain type, compounds of a certain type, can be subjected to a cluster analysis and, if a natural grouping is found, the resulting clusters correspond to a subclassification of the chemical systems which then can be utilized practically and/or interpreted theoretically.

9. Discussion

The analysis and interpretation of and complex relationships and the behavior of complex systems such as those investigated in Chemistry is often done using the concepts of similarity and classes. In the typical case one has observed relatively much data on each system (spectra, thermodynamic data, GC and trace element "spectra") but has relatively little theoretical information about the fundamental processes generating the data. By dividing the systems into classes so that the systems display a regular behavior (similarity) within each class, the analysis is considerably simplified; the complexity of a large collection of individual systems is reduced to the moderate complexity of a small number of classes.

The quantitative analysis of the data in each class by means of simple mathematical models such as eq. (1) makes it possible to extract most of the information contained in the data in a way which is easily interpreted. The linear structure of model (1) makes a graphical display of data and results straightforward.

What might be important in many applications is that SIMCA does not overestimate the structure within each class. The cross-validatory estimation of the number of components ($\underline{A}$) assures that only "real" regularities are described in the data. Thus, if there is no difference between two (or several) classes with respect to the data, this is found by the method and the chemist need not interpret artifacts introduced by the problem formulation.

It is philosophically interesting that the quantitative similarity model (1) seems to closely correspond to the intuitive meaning of similarity as the concept is used in chemistry. Thus, data observed on chemical systems which are said to be similar, can usually be described by model (1) with one term ($\underline{A}$=1).

When comparing SIMCA with other methods of quantifying similarity we note that SIMCA often performs better on real data sets that "optimal" methods such as the K-nearest neighbor method (5,30). The comparisons made on real data are too few to allow any definite conclusions, however.

Theoretically, SIMCA has the advantage before distance based methods of being only mildly dependent on the scaling of the individual variables. The independent handling of each class makes SIMCA insensitive to the number of classes which is a severely limiting factor with linear discriminant analysis and similar methods such as the linear learning machine.

Another severe limitation with most methods is that the number of variables must not approach the number of objects in one class. The number of available parameters then approaches the number of data with disastrous consequences for the robustness of the results.

SIMCA extracts only A components (eq. 1) from the data in each class and as long as A is small compared to the number of variables and objects (less than 1/4) in the classes, the results are robust even if the number of variables is very large.

Missing data are easily handled with SIMCA, a necessary condition for any method to be used in routine application.

We believe that SIMCA, being specifically developed for chemical data analysis, has many features which makes it useful for these purposes. So far, we have not detected any drawbacks with the method, it seems sufficiently flexible to adapt to most patterns revealed by real data. The experience with the method is still limited, however, and the application of the method to a large number of real data sets and the continuing comparison with other methods will provide further information about advantages and drawbacks of SIMCA in particular and quantitative similarity models in general.

Finally, we wish to stress that mathematical methods for data analysis such as SIMCA in no way are a substitute for chemical skill or intuition. Rather, analogously to a good instrumental method which simplifies the collection of accurate and relevant measurements, the mathematical method facilitates the extraction of relevant information from a given set of data. Thus, the chemist is freed from much of the drudgery of making lengthy calculations and can instead address himself to the more rewarding tasks of formulating problems, designing crucial experiments producing relevant data and interpreting the results of the data analysis.

Literature Cited

1. Theobald, D. W.; Chem. Soc. Rev. (1976), 5,203.
2. Schoenfeld, P.S. and DeVoe, J. R.; Anal. Chem. (1976) 48, 403 R.
3. Bard, Y. and Lapidus, L.; Catalysis Rev. (1968), 2, 67.
4. Sjöström, M. and Edlund, U.; J. Magn. Resonance (1977, 25, 285.
5. Wold, S.; Pattern Recognition (1976), 8, 127.
6. Kanal, L.; IEEE Trans. Inform. Theory (1974) 20, 697.
7. Cacoullos, T. (Ed.); "Discriminant Analysis and Applications", Academic Press, New York, 1973.

8. Jurs, P., and Isenhour, T.; "Chemical Applications of Pattern Recognition", Wiley, New York, 1975.
9. Kowalski, B. R.; in "Computers in Chemical and Biomedical Research" Vol. 2 (Ed.s, C. E. Klopfenstein and C. L. Wilkins), Academic Press, New York, 1974
10. Kowalski, B.R.; Anal. Chem. (1975) 47, 1152 A
11. Chapman, N. B. and Shorter, J.; (Ed.s) "Advances in Linear Free Energy Relationships", Vol. 1, Plenum Press, London, 1972.
12. Wold, S., and Sjöström, M.; in "Advances in Linear Free Energy Relationships" (Ed.s, N.B. Chapman and J. Shorter), Vol. 2, Plenum Press, London, 1978.
13. Bell, R. P., "The Proton in Chemistry", Methuen, London, 1959.
14. Hammett, L. P.; "Physical Organic Chemistry", 2nd ed., McGraw-Hill, New York, 1970.
15. Hammond, G. S.; J. Chem. Educ. (1974) 51, 558.
16. Anders, O. U.; Anal. Chem. (1972) 44, 1930.
17. Howery, D. G.; Int. Labor. (1976) March/April, 11.
18. Weiner, P. H. and Howery, D. G.; Anal. Chem. (1972) 44, 1189.
19. Malinowski, E. R., and Weiner, P. H.; "Factor Analysis in Chemistry", to be published.
20. Draper and Smith, "Applied Regression Analysis", Wiley, New York, 1966.
21. Wold, H.; in "Festschrift for Jerzy Neyman" (Ed., F. N. David), Wiley, New York, 1966.
22. Wold, H.; in "Multivariate Analysis", (Ed., P. R. Krishnaiah), Academic Press, New York, 1966.
23. Wold, S.; to be published.
24. Box, G. E. P. and Cox, D. R. J. Roy; Statist. Soc. B (1974) 36, 11.
25. Kowalski, B. R., J. Amer. Chem. Soc. (1973) 95, 686.
26. Wold, S., and Sjöström, M.; Chem. Scripta (1972), 2, 49.
27. Sjöström, M., and Wold, S.; Chem. Scripta (1974), 6, 114.
28. Sjöström, M., and Wold, S.; Chem. Scripta (1976), 9, 200.
29. Carey, R. N., Wold, S., and Westgard, J. O.; Anal. Chem. (1975) 47, 1824.
30. Duewer, D. L., Kowalski, B. R., and Schatzki, T. F.; Anal. Chem. (1975) 47, 1973.
31. Strouf, O., and Wold, S.; Acta Chem. Scand. A (1977), in press.
32. Wold, S.; J. Chromatogr. Sci. (1975), 13, 525.

33. Wold, H.; in "Perspectives in Probability and Statistics", (Ed., J. Gani) Academic Press, New York, 1975.
34. Wold, H.; in "Quantitative Socioloty" (Ed., H. M. Blalock), Academic Press, New York, 1975.

INDEX

INDEX

U

V

W

X

Z